1988

ICMI Study Series Editors A. G. Howson and J.-P.

School Mathematics in the 1990s

Prepared by
Geoffrey Howson and Bryan Wilson

and based on contributions by
Wasil Al-Dahir (Kuwait)
Marjorie Carss (Australia)
Ubiratan D'Ambrosio (Brazil)
Peter Damerow (German Federal Republic)
William Ebeid (Egypt)
Hiroshi Fujita (Japan)
Geoffrey Howson (England)
Edward Jacobsen (UNESCO)
Jean-Pierre Kahane (France)
Jeremy Kilpatrick (USA)
Emilio Lluis (Mexico)
Bienvenido Nebres (Philippines)
Stefan Turnau (Poland)
David Wheeler (Canada)
Bryan Wilson (England)

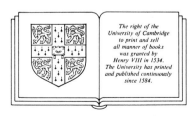

The right of the
University of Cambridge
to print and sell
all manner of books
was granted by
Henry VIII in 1534.
The University has printed
and published continuously
since 1584.

CAMBRIDGE UNIVERSITY PRESS
Cambridge
London New York New Rochelle
Melbourne Sydney

Published by the Press Syndicate of the University of Cambridge
The Pitt Building, Trumpington Street, Cambridge CB2 1RP
32 East 57th Street, New York, NY 10022, USA
10 Stamford Road, Oakleigh, Melbourne, 3166, Australia

First Published 1986

Printed in Great Britain at the University Press, Cambridge

Library of Congress cataloging in publication data available

British Library Cataloguing in publication data

School Mathematics in the 1990s. - (ICMI study series; v. 2)
1. Mathematics - study and teaching
I. Howson, A.G. II. Wilson, B. III. Series
510'.7'1 QA11

ISBN 0 521 33333 4 Hardcovers
ISBN 0 521 33614 7 Paperback

Contents

Foreword   v

Acknowledgements   viii

1. Mathematics in a Technological Society   1
2. Mathematics and General Educational Goals   7
3. The Place and Aims of Mathematics in Schools   19
4. The Content of the School Mathematics Curriculum   37
5. On Particular Content Issues   55
6. Classrooms and Teachers in the 1990s   75
7. Research   83
8. The Processes of Change   91
9. The Way Ahead   99
Bibliography   101

Foreword

The International Commission on Mathematical Instruction is planning a series of studies on topics of current interest within mathematics education. The first study was on the impact of computers and informatics on mathematics and its teaching at university and senior high school level. That study had as its centrepoint an international symposium held in Strasbourg, France in March, 1985 and attended by some seventy participants drawn from thirteen different countries.

The second study, which gave rise to this volume, has taken a slightly different form. First a discussion document, <u>School Mathematics in the 1990s</u> by A.G. Howson, B.F. Nebres and B.J. Wilson, was sent to all National Representatives of ICMI and circulated widely in the original English and in translation. A small, closed international seminar was then held in Kuwait in February, 1986 at which an invited group of mathematics educators, named on the title page of this book, considered issues raised in the discussion document, points made by those who had responded to that paper, and, of course, attempted to remedy its many omissions. This book is based on those discussions and has been prepared by Geoffrey Howson and Bryan Wilson. It is, as such, a compilation of views and certainly would not have the effect of drawing discussion amongst those who participated in Kuwait to an end. Its aim, however, is not to terminate discussions, but rather to provoke and stimulate them. Further serious and detailed debate will be required before sound responses to problems can be formulated. It is ICMI's hope that this book will facilitate such debate and decision-taking.

The subject is a vital one, for there are few issues which are of such concern to all countries throughout the world. Mathematics education faces a vast variety of challenges as we move towards the 1990s and it is by no means certain what responses to some particularly crucial questions will prove most appropriate. There are few certainties. Perhaps one of the few things of which we can be certain is that ICMI's approach to the problems must be based on a broader appreciation of mathematics education than is suggested by its somewhat outdated name. 'Instruction' or 'teaching' will always remain a key issue for us, but, as is emphasised in this volume, 'learning' demands equal consideration, and, what is still insufficiently recognised, this must include consideration of what mathematics students learn, and of the mathematical activities in which they engage, outside, as well as in, the classroom.

Foreword

　　We hope also that this book will prove timely. In many
countries there is a growing lack of confidence in our ability to teach
mathematics successfully. Education is poorly regarded. Such a crisis
of confidence must be overcome, for there is no doubt that
well-informed, active citizens in the 1990s will require more
mathematics and a greater comprehension of mathematics. We must react
against any tendencies to see the problems of mathematics education as
intractable.

　　Finally, I should like to acknowledge with gratitude the
leading role which Geoffrey Howson has played at all stages of this
study.

                                        Jean-Pierre Kahane

August, 1986.

Acknowledgments

The Kuwait Symposium of 1-6 February, 1986 on which this publication is based was financed and sponsored by the Kuwait Ministry of Education, Kuwait University and the Kuwait Foundation for the Advancement of Sciences, and was held under the patronage of Dr. Hassan Ali Al-Ebrahim, Minister of Education.

The International Commission on Mathematical Instruction is deeply grateful to all these bodies and to Dr. Al-Ebrahim for their most generous assistance and encouragement. It also wishes to place on record its indebtedness to the members of the local organizing committee, Mansour G. Hussein (Chairman), Dr. Adnan Hamoui, Adnan A. Al-Abdulmuhsen, Tahseen Budair and Mahmoud Abdul-Kadir, for their most excellent and greatly appreciated work.

We should also wish to express our thanks to the Institut für Didaktik der Mathematik, Bielefeld, where Geoffrey Howson worked on part of the manuscript of this book as a visiting professor, to Elizabeth Henderson, Roseanne Glover, Charles Jackson, Doug Jones, Kurt Killion, Kenneth Shaw and Siriporn Thipkong of Athens, Georgia, who offered detailed criticisms of a draft manuscript and many of whose suggestions have been incorporated in this final version, and to Mrs. Margaret Youngs who has so carefully typed this volume.

Chapter 1

Mathematics in a Technological Society

1. A New Revolution

The world is currently in the throes of a new technological
revolution which is having an impact on society at least as great as
the Industrial Revolution. Moreover, the speed at which the effects of
this 'Information Revolution' are being felt is considerably greater
than that of its predecessor, as is its range. There are now few, if
any, societies anywhere in the world that are unaffected.

It is this fact above all that underlay the decision of ICMI,
the International Commission on Mathematical Instruction, to mount the
study of which this present publication is the tangible outcome.

Much curriculum development in school mathematics over the past
thirty years has taken place in a piecemeal manner. Everyone
professionally concerned with mathematics education – and many who are
not, including many parents – has his or her own views on school
mathematics: what should be taught, how it should be taught, how fast
it can be learned, how it should be assessed .... . In order to
represent as wide a range of views as practicable, while still being
able to formulate guidelines sufficiently specific to be useful, a
group of experienced mathematics educators were invited to meet
together for six days in Kuwait in February 1986 to consider school
mathematics in the 1990s. These participants, drawn from six
continents, were selected as having a variety of personal viewpoints
and as representing different mathematical and educational traditions.
This publication is based on the discussions which took place at that
Kuwait meeting.

Many of the changes over the past thirty years have resulted
from curriculum developers in one educational system simply copying
what is being done in another. This partly explains the similarity of
mathematics curricula, particularly at secondary level, across the
world. The substance of these similar curricula is referred to in this
present study as the canonical mathematics curriculum.

There is no intention of prescribing a new orthodoxy. The
tenor of what follows is rather in the direction of greater diversity,
with mathematics curricula being developed with more attention to the

cultural circumstances and employment patterns of the country
concerned. Consequently, no answers are given to the issues raised.
The purpose is rather to identify some of these key issues which
mathematics curriculum developers will have to consider in the
immediate future. For some of these issues, various alternative
courses of action are suggested, and some of the consequences of each
alternative are mentioned. In this way it is hoped to initiate more
detailed local debate, country by country. This study offers a
framework within which such debate can take place in a coherent
fashion.

2.   Changing Demands

Through its influence on society, modern technology is causing,
and will increasingly cause, educational aims to be rethought.
Mathematics itself is being directly affected, as new branches are
being developed in response to technological need (see, for example,
ICMI 1986a), other time-hallowed techniques are falling into disuse,
and the balance of mathematical skills needed by the citizen to
function effectively in daily life is changing. In addition to this
direct effect, the teaching of mathematics is being affected in a
variety of other ways, through the changing demands, expectations and
employment patterns within society, through changing educational goals
and structures, and through changing pedagogical possibilities. The
canonical school mathematics curriculum and, indeed, state educational
systems, developed largely in response to the demands of the Industrial
Revolution; it seems self-evident that the Information Revolution will
result in major changes in both schools and their curricula, as new
demands are made and new opportunities provided for teaching and
learning in educational systems across the world.

Of course, the impact of technology is not the same in all
societies. For example, in rural societies within the Third World,
technological development can bring in its wake increased demands for
mathematics: subsistence and small-scale commercial farming can now
require mathematically-based decision-making, as well as the ability to
use machinery and to control it in response to readings on a variety of
scales and measures. How does the development of such abilities fit
into the school mathematics curriculum? Bearing in mind the high rate
of drop-out during the primary school years in many such countries (see
Figure 3.1), can the first few years of mathematics education be designe
in such a way as to make children functional in the processes they will
need in rural life? Further, would such a reordering of curriculum
priorities be consistent with the development of skills and attitudes
that would encourage pupils to continue to learn once they had left
school? Such questions will be taken up again in Chapter 3.

In technologically advanced countries, on the other hand, there
is evidence that some of the mathematical demands that used to be
placed on the bulk of those in employment are no longer required; many

traditional demands are nowadays met by the ubiquitous 'chip' (see, for example, Fitzgerald (1981) and note the report of the US Bureau of Labor Statistics (Romberg, 1984, p.6) that not only had most job openings in 1980 low skill requirements, but that this situation was likely to persist for the next quarter of a century). An area of life where such reduced demands are evident to practically every citizen of a technologically advanced country is that of shopping. No longer does a shop assistant have to be able to add up a bill, nor to subtract to calculate change, and neither does the customer have to check such arithmetic; a machine does it all. With the prospect of a cashless society, even the need to be able to count money could disappear. Yet as the flood of information available to every citizen rises both in employment and in daily life, other mathematical awarenesses are needed to handle it efficiently: ideas from statistics, probability, estimation, orders of magnitude, and understanding the assumptions that underlie a prediction or procedure. What is lessening is the need for particular skills, particularly of arithmetic; what is increasingly needed by all is an appreciation of more generalised mathematical concepts and ideas.

Such reducing demands on particular mathematical skills of many citizens, within societies in which mathematics itself is playing an ever increasing role, is a paradox. One result is that there is an employed élite on whom greater, and changing, mathematical demands are placed and who, as a result, are coming to enjoy increasing power. As in other aspects of life, there seems to be growing polarisation in the mathematical demands of employment. These factors must affect curriculum design, pointing to the need for greater diversification of curricula. There is now also the serious and growing problem of long-term unemployment to be considered. How should this and other factors affect the goals of 'mathematics for all'?

Technology, too, affects our ideas of what comprises 'useful knowledge'. Does the long-division algorithm remain in this category in a world in which there is ready access to calculators? Yet the concept of division retains its importance, and even the algorithm itself may still be 'useful' in a propadeutical sense to the minority of students who may later wish to factorise polynomials (unless they in their turn make use of computer software for symbolic manipulation). It must be borne in mind also that there are other varieties of 'usefulness' as well as immediate practical use (such as many specific skills concerned with sport, art, crafts, ...). There is knowledge of non-immediate practical use, where the learner is not yet in a position to need it (e.g. much of school 'commercial arithmetic', concerned with taxes, budgets, investments and the like); there is knowledge of more general, unspecific use (e.g. literacy and numeracy, enabling the learner to do a vast range of practical things); there is career-orientated individual usefulness (e.g. the calculus for potential engineers); there is usefulness for society (e.g. operational research, which enables certain kinds of important problems to be solved, though not everyone in the society needs to know how). All

these facets of usefulness must be considered in formulating an answer
to the question: What is likely to comprise 'useful' mathematical
knowledge in the 1990s?

Students' attitudes to schools and schooling, and their
conception of what comprises desirable knowledge and understanding, are
also affected by a technological environment. This is yet another
important development to which a response must be made.

Despite the considerable and growing impact of technology on
schools, our experience of all previous phases of educational change
leads us to believe that teachers will continue to play a central role
in students' education. However, the nature of that role is changing,
and will continue to do so. Technology challenges the present role
that most teachers have of being the chief source of knowledge for
their students. Children learn a great deal about their world from
television and other media, while there is much wider variety of, and
easier access to, print materials than during the teacher's own
childhood. This change of role from source of information to manager
of the curriculum will be a difficult one for the present generation of
teachers, and they will need much help and support in making it.
Technology can and will affect classroom practice: but in what ways?
Here we have a choice of approach. Either we can respond to the latest
developments – how do we use the micro? – or we can seek to influence
the designers and producers by setting educational goals for future
technology. What kind of classroom do we envisage for the 1990s? What
technological developments would we wish to take place? How can we
help fashion the development of technological hardware for educational
purposes? How may the teachers of the 1990s be prepared for new, and
to some extent unknown, developments through both pre-service and
in-service training? In later chapters we will consider these, among
other matters, in greater detail.

3. Mathematics Education in and for Society

It will be clear that technological developments, and responses
to them and to the social problems which accompany them, inform much of
this study. The general question therefore arises as to whether the
teaching of mathematics should be more explicitly related to some of
the social issues characteristic of our modern world. The particular
issues would vary from one country to another, but could include
matters like the statistical relationship between smoking and lung
cancer; the proportions of the national budget spent on defence, on
health and on education; the development of a critical attitude to
statistics put out both by commercial and by official sources; the
quantitative gulf which separates the rich and poor countries. The
discussion of this question illustrates the format which is used
throughout this study.

**Can we perceive a new social role for mathematics education in
a world in which technology plays a dominant role?**

Alternative 1.   Mathematics is neutral, and is best taught in
                        isolation from contentious social issues.

Consequences: 1. Teachers will continue to feel comfortable in
                          keeping strictly within the confines of their
                          own subject specialism.
                       2. The teaching of mathematics will continue to
                          be given high priority by all governments,
                          who see it as a crucial tool for economic and
                          technological advance.
                       3. Mathematics, in the eyes of the public at
                          large, will continue to retain its aura of
                          mystique and purity, above the common concerns
                          of mankind.
                       4. Mathematics education will make no direct
                          contribution to the urgent social issues of
                          this generation.

Alternative 2.   Since mathematics underpins both technology in
                        all its manifold forms, and the policies that
                        determine how it is used, its teaching should
                        deliberately be related to these issues.

Consequences: 1. It is very difficult to do. Indeed, many – if
                          not most – mathematics teachers will not see
                          it as part of their duty to touch on social
                          and contentious issues.
                       2. Governments are likely to respond adversely.
                          This has already happened in some countries
                          which have attempted to include a 'social
                          responsibility' component in the teaching of
                          physics, or to introduce 'Peace Studies' into
                          schools.
                       3. Student motivation is likely to be enhanced.
                       4. Mathematics educators might make a direct
                          professional contribution to some of the major
                          issues facing human society.

        (See, for example, Chapter 1 of Christiansen et al (1986) for a
more detailed discussion of possibilities and of difficulties.)

        It is hoped that the use of this 'alternatives and
consequences' format in subsequent chapters will facilitate
constructive discussion of the issues concerned. In considering them
in the light of their own circumstances, teachers, students, curriculum
developers, and other mathematics educators working in different
educational systems will – and should – sometimes come to different
conclusions. Often they may consider that a mix of two or more of the
alternatives set out is the most appropriate response. Such 'mixes'
are not usually included explicitly among the alternatives, since to

describe the extremes helps to clarify what it is that is being mixed.
As we have already indicated, our aim is not to promulgate any
particular doctrine, rather we seek to identify key issues on which
decisions will have to be made, and to suggest possible responses
together with probable consequences.

Chapter 2

Mathematics and General Educational Goals

1. Mathematics in the School Curriculum

Before trying to subdivide the problem area, it seems useful to consider some general issues that arise from the specific social contexts in which mathematics is taught, and the fact that mathematics is but one component of school life.

School systems are a relatively new phenomenon in historical terms, having only developed during the last hundred years or so. Before then, there were schools in some societies, but these tended to live independent lives united only by their religious underpinning, be it, for example, Christian or Moslem. Education, however, took place before there were schools, and much of what children learn is still learned outside school. Yet much of the socialisation that previously took place in and around the home now takes place in school, and it has now become the prerogative of the school to teach certain limited and specific skills and areas of knowledge. This range of knowledge and skills comprises the formal school curriculum, and contained within it is much of the mathematics that children learn.

How did mathematics come to achieve its central place in the school curriculum? Originally school systems offered education principally at what is now called primary (or elementary) level, and the secular curriculum was almost wholly devoted to the 'Three R's': Reading, Writing, Arithmetic. As the beginnings of secondary level education emerged, this curriculum was expanded to include language (both native and foreign), mathematics, science (physics, chemistry, biology), history and geography, and, later, art and sport. The basic, academic curriculum was proposed in England by the pioneering Chemist, Joseph Priestley, in 1760, and it is astonishing to realise that it is still the standard pattern across the world over 200 years later. The surprise is even greater when one considers what does not appear in it: medicine, economics, politics, technology, individual and group psychology (i.e. understanding people).... The persistence of this curriculum structure aptly illustrates the essential conservatism of the educational process, a conservatism equally evident when we consider the history of the mathematics curriculum.

The familiar school mathematics curriculum was developed in a particular historical and cultural context, that of Western Europe in the aftermath of the Industrial Revolution. Those who framed it only had a minority of society in mind, for at that time only a small élite sector had access to a substantial number of years of schooling. In recent decades, what was once provided for a few has now been made available to -indeed, forced upon - all. Furthermore, this same curriculum has been exported, and to a large extent voluntarily retained, by other countries across the world. The result is an astonishing uniformity of school mathematics curricula world-wide.

It is still true that, faced with a standard school mathematics textbook from an unspecified country, even internationally experienced mathematics educators find it almost impossible to say what part of the world it comes from without recourse to the essentially non-mathematical clues of language and of place-names.

If such uniformity was the result of the 'universality' of mathematics itself, it might be justified. Yet few would argue that mathematics holds its central place in school education simply for its own sake. Its justification as usually stated is to a large extent its 'usefulness', in employment and the future daily lives of students as citizens. Seen in this light, such uniformity is strange. Employment opportunities vary widely, making very different demands on both the nature and levels of mathematical skills and understanding, while the societies concerned span the range from subsistence living to high-technology urban life. There is urgent need for each country's curriculum developers to make a more radical assessment in order to determine a mathematics curriculum appropriate to the needs of their nation in the 1990s.

2. Mathematics for All

The movement towards 'mathematics for all' has given rise to major problems in mathematics education. It has happened so quickly that the full consequences have still to be evaluated and appreciated. New policies will be required for the 1990's. Much thought is already being given to what they should be (see, for example, Damerow et al (1986), Keitel (1985)).

Should mathematics remain at the heart of the school curriculum for all?

Alternative 1. No; 'real mathematics' cannot be taught to everybody.

Consequences: 1. Mathematics will no longer occupy a privileged position as part of the core of general education.

                 2. An academic élite will study 'real mathematics'.

3. The majority of students will meet only 'useful mathematics', and this in timetable slots labelled physics, technical education, economics, etc.

4. Problems of course selection for different students arise.

Alternative 2. Yes; mathematics must be planned so that it can be effectively taught to all.

Consequences:
1. Mathematics will appear to retain its central place in the school curriculum.
2. This new school mathematics might differ substantially from that traditionally taught.
3. The gap between school mathematics and higher mathematics will grow.
4. Equality of opportunity for all students will be seen to be preserved.

Alternative 3. Yes; but it is accepted that although taught to all it will not be understood by all.

Consequences:
1. Mathematics will retain its status both in the school curriculum and in popular estimation.
2. All students who are capable of understanding the mathematics in the curriculum will have the opportunity to do so.
3. Many students will, as in the past, experience frustration and discouragement.
4. Much teacher time will be wasted in teaching a style of mathematics to students who have effectively given up on the subject.

Alternative 4. Yes; but students will be taught different types of mathematics, or will be taught the same mathematics at different rates, according to their levels of 'ability' or standards of 'attainment'.

Consequences:
1. Mathematics will retain its place in the core of the school curriculum.
2. All students will have opportunity to study a style of mathematics appropriate to them as individuals.
3. Problems of course selection for different students arise.
4. Curriculum planning and the provision of appropriate resources becomes very much more difficult and expensive, at all levels, than with a uniform curriculum.

3. Differentiation of Students and Curricula

Of the alternatives given in the previous section, it is the last, (4), which currently would appear to attract most support. But how can effective differentiation of students and curricula be achieved? To what degree is differentiation desirable/possible?

The range of achievement among different individuals exposed to the same opportunities is greater in mathematics than in almost any other aspect of intellectual endeavour, with the possible exception of music. Such self-evident variation in mathematical capability within a population leads naturally to demands that different mathematical diets be available at school. Yet nearly all attempts to introduce effective differentiation of curricula have failed. Various possible strategies for such differentiation will be considered later, in the hope that they might point the way forward to some resolution of this major current problem in mathematics education.

This lack of success in introducing effective differentiation of mathematics curricula is particularly acute in the developing countries, where the waste of financial and human resources represented by failure at school can be least afforded. This results in increasing failure rates in public examinations as the proportion of the nation's children who go to secondary school rapidly increases, while the school mathematics curriculum remains essentially unchanged from that which was planned for a small academic élite in earlier years.

The reasons for this state of affairs are almost all social and political. While there would be widespread agreement among educational planners to the proposition that differentiation is needed, decisions as to which child follows which curriculum are taken individually, and every individual pupil - and parent - insists on doing the course that opens the way to the greatest opportunities. This multitude of individual decisions is founded on a sense of equal opportunity, which is as strongly developed in a young nation as in those older countries in which egalitarianism has been an increasingly powerful force in recent years. Yet the demand that educational provision shall be equitable for all may well be in conflict with the other widely-held belief that education should allow each individual to realise his or her maximum potential.

Attempts to differentiate by institution have been no more successful than schemes to differentiate within institutions; politicians will rarely risk the unpopularity of having second-class schools established within their constituencies. Can a way be found to meet the widely differing mathematical needs of students at school, while circumventing these social and political objections to overt differentiation? Can it be done without continuing to pay the unacceptable price of widespread 'failure'? The issue is fundamental to many of the decisions that have to be taken by curriculum developers. It will be discussed further in the next chapter in the context of the position of mathematics in the total school curriculum, and it will be considered in alternatives-and-consequences format in Chapter 4.

4.   The Special Place of Mathematics in the Curriculum

A major reason for the persistence of the special place held by mathematics in the school curriculum is the way in which it has been used in the last two centuries as a screening device, or 'filter', for entry to numerous professions. In this role it replaced Latin, which consequently all but disappeared from schools. Would the same happen to mathematics if an alternative filter mechanism were introduced? It hardly seems likely. So many parties view it as being of major importance, though not necessarily for its own sake. It is salutary to remember that it is the only subject taught in practically every school in the world. Yet the very esteem in which it is held may militate against significant change and the subject's well-being, since people hesitate to initiate fundamental change in something that is already highly regarded in its present state.

Should the privileged position of mathematics in school be challenged or should mathematics educators fight to uphold it? Would school mathematics better serve the needs of more students if its position in the curriculum were revalued, for example by devoting less (or even more) time to its study? (We note here recent changes in the French 'bac', as a result of which less time is devoted to mathematics teaching for several groups of students.) In what ways can mathematics contribute to general educational goals? Let us consider four non-exhaustive aspects of that contribution: the development of reasoning power, its exemplification of certainty, the aesthetic pleasure it provides, and its service role to other disciplines (other views on how learning mathematics can serve to develop "basic academic competencies" can be found in College Board (1985)).

5.   The Contribution of Mathematics to Education

For many centuries mathematics was seen as the subject par excellence in which 'reasoning' powers could be trained. The most common popular answer to the question as to why there is so much mathematics in school was - and often still is - "it teaches you to think". It is, as the sixteenth century Spaniard, Vives, put it, a subject 'to display the sharpness of the mind'. This was indeed seen as the justification for the pre-eminent place held by Euclid's geometry which is one of the abiding memories of secondary school for generations of people across the world. Yet there is little evidence that school mathematics in fact possesses the ability to develop reasoning outside itself, and few would now make claims for it similar to those made by nineteenth-century educators. Perhaps 'reasoning powers' has been replaced by the development of 'critical powers', that a mathematical education enables people to handle the mass of data with which they are constantly bombarded in this information age. An implication would be that simple statistical ideas should be part of every secondary school student's education, to help to develop a critical attitude to numerical information presented by the mass media.

A major characteristic of our age is a loss of certainty, affecting almost all aspects of human experience: in politics, in religion, in economics, in the arts, in the understandings of science, in the future of civilization itself. The restructuring of traditional patterns of social and family life across much of the world, in both the industrialised and the developing countries, leaves children and adults with a sense of insecurity and uncertainty more widespread than in any previous generation. For many children, school represents a stability unknown elsewhere. Yet within the school curriculum, children realise that much of what they do is judged by opinion, whether it be the quality of an essay, a painting or an attempt to pronounce a foreign language. Even an apparently factual subject such as history has to be taken on trust. Only in mathematics is there verifiable certainty. Tell a primary child that World War 2 lasted for ten years, and he will believe it; tell him that two fours are ten, and there will be an argument. Children know what is right and what is wrong at their own level of competence in mathematics, and can verify it themselves, even though they may not always be encouraged to do so. "Thanks to mathematics, man is able to be sure of something" (Alberto Barajas), and, as a consequence, "Mathematics has produced for humanity an immeasurable psychological well-being. We are no longer afraid of insane gods playing merciless games with us human beings". Such certainty that there exists something that is true, beyond doubt, something in which all can come to total agreement, is everyone's birthright. There are at least two implications for school mathematics. First, every student must be enabled to experience enough mathematics to be convinced that there is something that is true, beyond any doubt. Secondly, the teaching style should be such as to encourage and enable students to come to such convictions for themselves.

All who have ever produced mathematics at a level original to themselves know the pleasure that it can give. Creating a positive attitude towards the subject is one common aim of school mathematics, and helping students to experience such intellectual pleasure is a means to that end. Yet in practice such pleasure is restricted to far too few, at any school level. Perhaps the trend towards more open-ended investigations in the mathematics classroom will enable more students to have a first experience of the aesthetic rewards that the Queen of the Sciences affords her subjects.

The service role of mathematics is of ever-increasing importance. This is a separate issue from the 'filter' role of the subject, since many professions which require evidence of mathematical attainment for entry do not make use of the mathematical skills and knowledge demanded in the qualifying process. The effects of its service role on the teaching of mathematics are, in fact, mixed. On the one hand, subjects such as biology and geography are becoming more quantitative and are providing new opportunities for coordinating school work and for displaying the power of mathematical applications. On the other hand, as the result of the need to teach physics to many more students, the mathematics traditionally found in physics syllabuses has often been weakened, since traditional demands cannot be met by the majority of students; qualitative arguments have been given greater emphasis, and there has been a corresponding lack of reinforcement of what has been taught in mathematics lessons. The service role of mathematics at tertiary level is the subject of a separate ICMI Study (ICMI, 1986b).

6. Mathematics and Difficulty

By and large, people regard mathematics as a hard subject. For many, it is associated with a strong sense of failure, and their memories of school mathematics are of tests and examinations, of crosses, of the fear of "getting it wrong" (Buxton, 1981). It was the subject above all which sorted out the academically bright from the dull. In view of the way in which so much mathematics has been taught in the past, this is not surprising. More perversely, however, there is a widespread popular feeling in some countries that school mathematics should be difficult, a feeling perhaps associated with a vague belief that it has a role in character-training. Such a belief runs strongly counter to the efforts of mathematics educators to raise the quality of mathematics teaching and learning, and to make the subject accessible and more enjoyable to a greater proportion of children. It is a belief that must be countered by every means possible, and it underlies, for example, the extraordinary reluctance of many teachers, even whole education systems, to allow calculators into schools within societies where they are readily available outside schools to all who have need of them.

7. Mathematics and Memory

Another issue which modern technology is forcing us to reconsider is the role of memory in education, and in particular in mathematics education. In a pre-literate society, memory was the repository of all that society's inherited knowledge and skills. Literacy meant that the burden could be shared between the remembered and the written, but it was only the invention of printing that enabled the written record to become widely available. Education, as the process of transmitting that part of a society's inheritance it considered important to make available to the next generation, responded by making learning to read central to schooling. Memory, however, continued to play a substantial role in school learning. Until recent years one major education system insisted that all its primary children learned the multiplication table up to 47 x 47. The current information revolution makes available virtually unlimited long-term memory in electronic form, access to which will become increasingly available in both homes and places of work. This raises the question

of whether there is an irreducible core of information which it is
necessary for individuals to remember, rather than to know how to
access when needed. If so, what is that core for mathematics? Who
decides? Does the availability of electronic memory do anything for
school mathematics that printing did not? Is there a danger that this
availability will result in greater emphasis on the 'transmission'
aspects of mathematics education at the expense of process and
exploration aspects? These are the kinds of question that must be
answered for each educational system in the light of its own
circumstances. Ideally, educational practice will thus respond to this
aspect of technological advance. However, the persistence of the
traditional university lecture half a millenium after the invention of
printing gives one cause to wonder whether practice always responds to
technological change as quickly as it might.

8. Special Needs of the Developing Countries

        The need to reconsider and make adjustments to the traditional
mathematics curriculum (a phrase used in this context to embrace the
'new math' curriculum) becomes even more apparent when one considers
the needs of developing countries. In many, the educational systems
have arisen as clones of those in the countries of the former colonial
powers, and, once established, prove remarkably difficult to change in
any significant way. Originally secondary education was exceedingly
limited in its availability, seeking to offer to a minute élite an
education comparable to that offered in the educator's home country and
which had itself been designed as a basis for university education in
Western Europe. Such an aim no longer makes sense, in educational, in
social or in political terms. Its result is to sacrifice the needs of
the many to the perceived needs of the few -'perceived' because in
reality the particular needs of the few who proceed to university
education could be met by an interpolated period of special study, as
is already happening, for example, in some African countries prior to
students proceeding to university in their own country. The present
situation is that 'comparability' has taken precedence over
'desirability', to the detriment of the education of vast numbers of
children in country after country of the developing world. Ironically,
it is no longer even a contemporary comparability; educational practice
in the former imperial countries has itself changed dramatically, with
major efforts being made to diversify curricula to meet the real, and
very varying, needs of the secondary school population. Attitudes to
'schooling' in these latter countries are also changing, with many
signs that schooling is held in less regard than it once was, a change
accentuated by technological developments which mean that entertainment
and non-formal educational means - particularly television - are of a
quality that cannot be matched by conventional school approaches. This
theme is taken up again in Chapter 6.

        The socio-economic barriers which stand in the way of
improvements in the developed world are also, of course, much more
pronounced in the Third World. Facilities are more limited; sometimes

the same school buildings have to be used by two or even three shifts
of students each day, with consequent restrictions on classroom hours
and on the possibility of preparation in classrooms or in laboratories.
Teachers are inadequately trained, overworked, poorly paid and held in
lower esteem than was once the case. Textbooks are in short supply,
and where they exist have probably been written for students having
different needs. How can educational advances be planned which take
into account such constraints? The creative use of television via
satellite transmission for educational purposes in India shows what can
be done even with limited resources.

9. Ethnomathematics

Sadly, the mathematical knowledge, ideas and intuition with
which children arrive at school, derived from pre-school experience
within their own environment, are largely ignored. Children first
coming to school are treated as tabula rasa, to be taught a
pre-determined programme of mathematics from the beginning. Yet in any
socio-cultural group there exists a great variety of tools for
classifying, ordering, quantifying, measuring, comparing, dealing with
spatial orientation, perceiving time and planning activities, logical
reasoning, relating events or objects together, inferring, acting with
regard to existing facilities, dependencies and restraints, and so on.
Although these are mathematical activities, the tools are not usually
explicitly mathematical tools. However, they constitute the basic
components of mathematical behaviour, the development of which must
surely be a major objective of school mathematics teaching. The
manipulation of these tools with a clearly defined objective or
intention is the result of recognisable patterns of thought, rather
than merely ad hoc practices. This complex of thought-patterns and
systematic practice has been termed the 'ethnomathematics' of the
cultural group concerned (D'Ambrosio, 1985, 1986). The mathematics
with which a child first comes to school will contain elements of such
ethnomathematics. Yet in traditional primary education, such
ethnomathematical knowledge is largely ignored; this is in sharp
contrast to, for example, the teaching of language, where the primary
teacher deliberately uses what children know and feel in order to
develop their linguistic skills. In mathematics, however, the
assumption is that the only mathematics that children know is what they
have learned in school. Such an assumption is similar to making
children learn in a language other than their own, a practice still
current in some educational systems.

Should school mathematics allow for children's
ethnomathematical knowledge?

Alternative 1.   Yes

Consequences: 1. Self-confidence is reinforced.
              2. Cultural values are respected.
              3. Mathematical activities remain spontaneous, and more
                  profitable for the majority of students.
              4. Teachers have to be aware of the cultural backgrounds
                  of their students.
              5. Teachers must be professionally and psychologically
                  prepared to listen carefully to students, and to
                  allow them to bring to situations under discussion
                  their own approaches on how to deal with them.
              6. The early primary mathematics curriculum must be
                  constructed with greater flexibility without laying
                  down a detailed list of topics and skills that must
                  be covered by specified times.
              7. Teachers must have some space to act as their own
                  curriculum builders, and in their training must be
                  given help in the principles of such curriculum
                  building.
              8. New textbooks and teaching aids with an
                  ethnomathematical orientation will be needed.

Alternative 2.   No.

Consequences: 1. The dichotomy in the student's mind between home and
                  school is reinforced.
              2. Cultural values are disregarded, and teachers can
                  teach mathematics effectively without any substantial
                  knowledge of their students' cultural backgrounds.
              3. Mathematical activities are seen largely as following
                  standard, teacher-taught paths.
              4. The mathematics curriculum can be predetermined from
                  the start of primary school.
              5. Traditional textbooks suffice.

An ethnomathematical example is a consideration of celestial
phenomena through naked-eye astronomy, and its relation to time
reckoning, to the calendar, to geometry, to astrology, to orientation,
to navigation, to the seasons, and so on. For large numbers of people,
whether from developed or developing countries, ethnomathematics is
essentially what they need and use for their entire lives; it is now
well known, for example, that very few people use the standard school
algorithms when faced with an arithmetical calculation in daily life
(see, for example, Carraher, Carraher and Schliemann, in Damerow et al,
1986). The dynamical character of ethnomathematical knowledge itself
provides for the development of further knowledge as it is needed; this
is an embodiment of the widely-recognised educational objective of
'learning how to learn'.

There remains, of course, the problem of transition from ethnomathematics to more formal mathematics. Teachers will need to be trained in the art of facilitating this transition, building on the potential mathematical creativity intrinsic to ethnomathematics. Very little work has as yet been done on this problem of transition.

There is a related need to introduce 'the mathematics of phenomena' into schooling. The demystification of natural phenomena will help to avoid ideological manipulation: Halley's Comet must not be viewed as the announcement of some disaster or good fortune. Another, and more positive, example is that political participation is essential in modern society. To participate fully in the political process every citizen needs some basic knowledge of statistics, economics and perhaps conflict situations (games theory). How may such new topics best be introduced into the school mathematics curriculum?

It is in the developing countries, too, that problems of language and mathematics education are most evident. Important questions arise relating to the language used as the medium of education. At what age, if any, should the medium change from the mother and/or national language to, say, English or French? Such evidence as there is points to the advantage, as measured in terms of mathematical understanding, of retaining the mother tongue at least throughout primary schooling (Fafunwa, 1975). Yet in practice such decisions are rarely made on educational grounds, since they involve more strident social and political considerations. Mathematics educators are usually left to make the best of policies decided on other grounds. They have some experience from other countries to draw upon; for example, Malaysia and Tanzania have in recent years pursued radical new policies in terms of the medium of school education, while countries such as Hungary and the USSR have long traditions of coping with linguistic minorities through the provision of texts in the mother tongue.

In some ways, being educated in a complex linguistic environment can be an asset in the development of mathematical understanding. Yet all too often in places such as North America or Western Europe, linguistic minorities are also socially deprived and this compounds their learning problems. What seems essential to mathematical progress is a sound grasp of, and facility in, one language (Dawe, 1983). Thus, maintenance of the first language should be encouraged not only on cultural grounds, but as a strategy to improve mathematical learning. This is not to imply that the first language should necessarily be used as the medium for mathematics education, but that it should be maintained in its own right.

Language issues permeate much mathematics teaching and learning even in a monoglot environment. This is a large and growing field of concern; the reader is referred to Jacobsen and Dawe (1986) and Wilson (1981).

10. Girls and Mathematics

        In this chapter we have not distinguished between the aims of a
mathematical education for girls and those for boys, for, in our view,
there are no essential differences. Yet there is no doubt that in many
countries enormous gender differences exist in terms of expectation and
achievement: a form of differentiation based on gender seems to be
occurring. What can be done in these countries to remove undesirable
gender differentiation as measured by attainment and participation in
occupations and further education courses for which mathematical entry
qualifications are required? Will the impact of technology in schools
tend to exacerbate the position, through role-assumptions and a
tendency to see computing and technical education in general as male
preserves? It is not proving easy to correct the gender biases
traditionally associated with mathematics education. Yet it is a
problem now being tackled with increasing vigour. (See, for example,
Becker and Barnes (1986), Royal Society (1986).) Like so many others,
it is a problem which greatly varies in extent from country to country.
It is essential, however, both for the well-being of mathematics in
school and more generally (because of the way in which school
mathematics acts as a filter and a gateway) for society in general,
that due attention is paid to eliminating gender disadvantage.

Chapter 3

The Place and Aims of Mathematics in Schools

1. The Canonical Curriculum

We have seen that the canonical school mathematics curriculum
was developed in Western Europe in the aftermath of the Industrial
Revolution, and has been adopted practically everywhere during the
present century. The importance with which the subject is invested
world-wide gives it an unrivalled position in school education. This
universal status, and the extraordinary uniformity of syllabuses across
the world, seriously inhibit significant changes in school mathematics
in any particular country. Yet it is this same status that means that
the teaching of mathematics absorbs a large proportion of the resources
of every education system - resources of finance, of teachers, of time.
Accordingly it is of paramount importance to ask fundamental questions
about the place and aims of mathematics in schools. Where social,
economic, cultural and employment patterns differ so dramatically
between countries, it is difficult to believe that everyone's needs are
being best served by the remarkable similarity of the ways in which
this central subject is incorporated into school curricula throughout
the world.

To give one example of this variety of socio-economic
circumstances, consider the school enrolment patterns of Mexico and of
Japan. Both countries have a six-year primary cycle followed by a six-
year secondary cycle (which in Japan is split into two three-year
phases, junior and senior). Both countries aspire to universal
secondary education for all. Yet, as Figure 3.1   shows, there are
great differences in the percentages of the age cohorts actually
undertaking full-time study. In practice, only about 60% of Mexican
children who start primary school continue beyond the first year.
Roughly 10% start secondary education, and a mere 3% complete it. By
contrast, practically all Japanese children complete lower secondary
education, and about 95% stay in full-time education until they are at
least 18. Yet the mathematics syllabuses year by year of the two
countries are very similar. Both are based on syllabuses developed
elsewhere, for pupils whose circumstances resemble those of students
neither in Japan nor in Mexico.

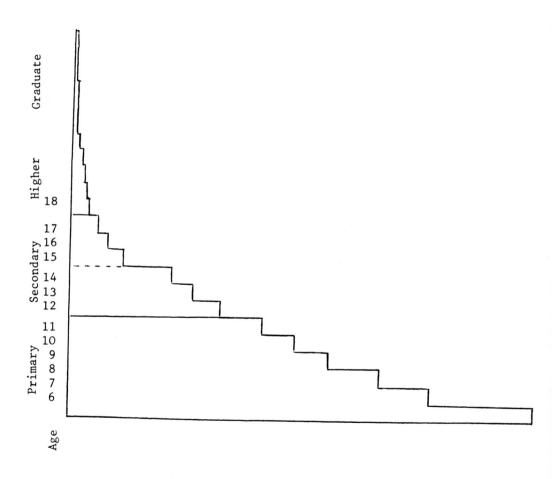

Fig. 3.1 The 1980 education 'pyramid' in Mexico. In contrast, about
          95% of Japanese 17 year-olds now receive full-time education.

This is not to imply that international exchange of
information, of ideas, of research findings and even of materials
within mathematics education is of no value. Every nation must want
the best for its children, and things are unlikely to improve country
by country in isolation. What is necessary, however, is that those
responsible for the development of school mathematics should become
much more critical of experience derived from elsewhere, and pay more
attention to their own actual circumstances and needs than to
considerations of international 'standards' and issues of
comparability. A consequence is that the most appropriate answers to
the questions raised in this present study will differ from country to
country, since the goals of school mathematics will not be identical
everywhere.

2.   The Aims of the Curriculum

Clearly, school mathematics must have a strong 'service'
element; it must serve to equip students both to study other subjects
and also to help cope with mathematical demands and problems which they
will meet out of school. (It must be emphasised that few 'real-life'
problems can actually be translated into purely mathematical terms;
there are usually qualitative judgements concerning taste,
desirability, pride, ..., as well as subjective probabilistic arguments
involved which lift such questions as 'Can I afford to ...?', 'Is it
time to trade in ...?', 'Which should I buy?' out of the realms of
mathematics per se. Indeed, this raises the question of whether or not
the utility of mathematics is best demonstrated in mathematics lessons,
in the form in which they currently exist. For even if such questions
cannot be answered entirely within the field of mathematics, the
subject does have an important contribution to make to their
resolution.) There are, therefore, certain 'proficiencies', such as an
understanding of percentages, which it is reasonable to ask school
mathematics courses to provide.

Is it then the purpose of school mathematics to formalise and
explain ethnomathematics and to provide techniques for problem-solving?

If one considers the three 'types' of mathematics

    (i)   ethnomathematics,
    (ii)   school mathematics,
    (iii)   higher (pure) mathematics,

then it can be argued that the reforms of the 1960s sought to
link (ii) and (iii) more closely: to transplant the aims, methods and
structures of (iii) to (ii). Now, greater efforts are being made to
link (i) and (ii). Yet, can school mathematics be divorced entirely
from (iii), other than as a provider of techniques? Should it not also
provide a foretaste of what mathematicians do, of why they do it, and
of the pleasure which the success of solving a mathematical problem can
bring?

It is not by chance that mathematicians tend to mature and achieve fame earlier than, say, biologists or historians. For a student can operate in much the same way as a professional mathematician far earlier than is the case for biologists or historians. It is not merely a case that greater knowledge is required to do original work in biology or history; acquiring the necessary methodology demands a greater maturity and a longer apprenticeship. The school student <u>can</u> act as a mathematician. This is demonstrated in a dramatic way by contestants in International Mathematical Olympiads, but it can also be shown by far more typical students when they are given the opportunity.

In the IMOs the participants act as problem-solvers, and it is in this area that outstanding talent is most readily displayed. The ability to acquire and restructure mathematical knowledge, to see new order and connections in what were formerly disparate parts of mathematics, is a different kind of mathematical talent, as is that of clear exposition. Yet attention can still be given at school level to these aspects of mathematics; students can be alerted to, and given examples of, the way in which mathematicians search for common structures, and they can be encouraged to record their mathematics and develop the skills of writing and talking about the subject.

Student motivation to study mathematics, then, will hinge on two aspects: 'profit' and 'pleasure'. At the moment, the profit to be derived from studying mathematics is too often thought of in an extra-mathematical sense: learning mathematics can be seen as a means of satisfying 'entry' requirements, of gaining necessary qualifications, of opening up the job-market (see Mellin-Olsen, 1981). This is a legitimate type of motivation to use, but if it is the sole motivation then it will seriously distort mathematics teaching. 'Profit' must be tied with more directly mathematical ends: students must appreciate that with mathematical knowledge and understanding they acquire desirable power. For they must learn that mathematics can help in the solution of <u>their</u> problems and in their own decision making. Pleasure, alas, is too infrequently associated with school mathematics, or even with schools. When it does occur, it is often the pleasure that the student gets when (s)he has grasped a technique, when (s)he can readily acquire ticks, marks and the teacher's favour. Again, this is justifiable and we must always remember the pleasure that 'success' - be it measured in even the simplest terms - brings, but one must try to develop other deeper, more aesthetic pleasure of the kind referred to in the previous chapter. Moreover, 'pleasure' must not only be provided for students; it is equally important that curricula are such as to encourage teachers to obtain 'pleasure' from their work.

3. The Compartmentalised Curriculum

        At the time when the outline of the standard school curriculum
was being formulated, knowledge was usually thought of as falling into
distinct and largely self-contained compartments, some of which became
school "subjects". This compartmentalisation of knowledge is the basis
of secondary schooling, and often also of primary schooling, in almost
every community. It has been effective for two reasons:

(i)   any human activity needs a structure, and the subject-based
          curriculum - which owes more to tradition than to logic
          -provides a convenient model for this purpose;

(ii)   non-vocational education has been the privilege largely of a
          section of the population endowed with high intelligence, and
          such individuals are able to create for themselves the
          connections between knowledge derived from different sources
          (this does not, of course, imply that they necessarily do so).

        Experience suggests that, in a system of education 'for all',
these conditions no longer hold. First, the conventional academic
structure is not seen as relevant to pupils' perceived needs, so that
motivation is weakened. Secondly, many pupils are unable to transfer
knowledge acquired in the isolation of an academic subject into usable
forms in the world outside school. The resultant frustration
contributes to low achievers' sense of alienation from school and all
that it represents during their later years in compulsory education.

        It seems clear that, for many pupils, a different model of
education is needed, and that structures need to be found which do not
involve this abstraction of experience into compartments, but which
provide vehicles for relating knowledge as it is acquired. Ideally,
the whole curriculum should be reviewed with this in mind. However,
the aims of this present study must be more limited and be based on the
non-controversial assumption that the overall school curriculum in the
1990s is unlikely to differ substantially from that we know.
Traditional subject 'boxes', including one labelled "mathematics", are
almost certain to be retained, though their boundaries are likely to
become increasingly blurred through the use of generic packages for
wordprocessing, spreadsheets, graphics and so on which focus on ways of
learning and handling information common to a number of current
'boxes'.

        How can the mathematics curriculum be developed so as to assist
students to relate the knowledge they acquire through its study to
other aspects of their experience?

Alternative 1.   Within a conventional topic-based syllabus,
                        teachers should make more effort to point out
                        relationships between different mathematical
                        topics.

Consequences: 1.   The subject would become more coherent to some
                       students.
                 2.   Little would be achieved in terms of relevance
                       to the rest of the student's experience.
                 3.   Only minor changes would be required of
                       teachers.

Alternative 2.   As each mathematical topic is taught, students
                       must have opportunities to solve problems that
                       demonstrate some of its applications in other
                       subjects and situations.

Consequences: 1.   Mathematics would be seen by more students to
                       be meaningful and of significance outside
                       itself.
                 2.   Most teachers would find greater demands
                       placed on their own knowledge and understand-
                       ing of the uses of mathematics in the world
                       about them.
                 3.   It would require closer co-operation between
                       teachers of mathematics and teachers of other
                       school subjects, particularly the sciences and
                       social sciences.

Alternative 3.   The curriculum should be restructured on the
                       basis of the applications of mathematics.

Consequences: 1.   Inter-disciplinary work would be encouraged in
                       schools; this could not come about by the
                       decisions of mathematics teachers alone.
                 2.   Major reorientation of the perspectives and
                       skills and understandings of many mathematics
                       teachers would be required.
                 3.   It would lead to more open-ended teaching and
                       learning, as problems would be discussed to
                       which there are no clear or unique answers.
                 4.   Mathematics could come to be seen only in its
                       Servant, not in its Queenly, role. In
                       particular, there would be few incentives to
                       abstract and generalise.
                 5.   Within such a system it would be difficult to
                       arrange for the practice and reinforcement of
                       mathematical techniques.

Alternative 4.   The curriculum should be restructured on the
                       basis of mathematical processes rather than on
                       the present basis of subject-content.

Consequences: This most radical alternative would require
                             the re-development of the mathematics
                             curriculum from the beginning, with new
                             sources of teaching material and a
                             reorientation of teachers' perspectives and
                             skills even greater than in the previous
                             alternative. Moreover, society's (parents,
                             employers, ...) views on school mathematics
                             would have to be re-adjusted.

     Some mix of the above alternatives is practicable, and probably
desirable. The optimum mix is a function of the particular education
system concerned, and in particular of the ability of the bulk of its
teachers to make the necessary adaptations.

4. The Process-based Curriculum

     The theme of applications of mathematics will be taken up again
in Chapter 5. Meanwhile, some further discussion of the last of the
above 'alternatives' is necessary. Traditionally, mathematics
curriculum-building has assumed that mathematics is a body of
knowledge, and that the mathematics educator's task is to take from it
a selection of topics that will best prepare students for their future
lives. However, an alternative view is possible. This sees
mathematics as a set of processes, the task of the school then being to
help students to learn how to 'mathematize'. Then, instead of deciding
which mathematical topics are indispensable for school leavers, the
task is to choose which processes might best serve a nation's citizens,
and then what experiences in school might help students to learn these
processes.

     Within a computer-orientated society, such processes would
include comparing, classifying, ordering, abstracting, symbolising,
generalising .... all of which may be subsumed within the term
mathematizing. Most people have some innate ability to do these
things; indeed, they permeate much of ethnomathematics. How can such
abilities be developed? Various other questions arise. What is the
relationship between the development of the ability to mathematize, and
the acquisition of mathematical knowledge? Is 'process' a practicable
basis for school curriculum-building? Would it be an acceptable basis
to a Ministry of Education and a public conditioned to think of
mathematics in terms of content and technique? Would it be welcome to
teachers? What are the major implications for student assessment
procedures? (See Chapter 6.) What kind of problem situations and
teaching materials can best assist the student to develop the ability
to mathematize?

     Processes can only be taught through content and so appropriate
mathematical topics will have to be taught in order to optimise the
chances of the process being understood and acquired. Will it be
possible to keep the emphasis on the processes, or is it inevitable

that the teachers' and examiners' emphasis will gradually become
focussed on the content and procedures, that the 'means' will be seen
as the 'end'?

The difficulties are formidable. Yet there is widespread
agreement among mathematics educators that greater emphasis on
mathematical processes is desirable at all educational levels.

Before leaving the question of the relationship of mathematics
to other school subjects, it should be pointed out that there may be
special features of mathematics, and of the problems of teaching and
learning it, that justify its standing apart as a separate subject in
the school curriculum. As was remarked above, the acquisition of
mathematical techniques requires practice and continued and planned
reinforcements. 'Understanding' would seem usually to be acquired as a
result of engaging in varied and repetitive mathematical activities. If
mathematics is only dealt with en passant in the learning of other
subjects, then opportunities to acquire techniques and to gain
understanding (which usually takes some time and rarely follows the
first meeting with a concept) will be very limited.

5. The Compulsory Curriculum

One result of the high status of mathematics is that in
practically all countries it is a compulsory subject for all students,
at least until the statutory minimum school-leaving age. In the USSR
it is a compulsory subject throughout schooling; so it is in Japan,
though it shares this with all other subjects, the school curriculum
being the same in broad outline for all students. In the United
Kingdom mathematics is in practice taken by all students throughout the
period of compulsory education, though there is no statute to say that
it must be. In some other countries, e.g. USA, the position is less
clear-cut, with greater freedom for individual students to continue
with mathematics or not as they wish, at least after a certain age. The
issue only concerns the secondary level of education; mathematics is
compulsory throughout primary schooling everywhere.

The pros and cons of making mathematics a compulsory subject of
school education have been argued for centuries, and will not be
repeated in detail here; they cover grounds of the utility of the
subject, its importance as an academic discipline, its contribution to
other disciplines, the experiences of creativity, enjoyment and beauty
that it can afford, its role in developing reasoning powers, and so on.
There are career arguments: to make mathematics optional can be a means
of allowing students seriously to limit their future employment
prospects. There are arguments from social ethics: in the interests
of equity, students should not be allowed to handicap themselves within
a competitive society by prematurely abandoning such a crucial subject
of study.

On the other hand, the consequences of opting for compulsory mathematics are clearly considerable. First, it increases the number of mathematics teachers needed, and means that many of them - and they are a precious resource in any education system - have to spend time teaching students who apparently gain very little from their efforts, at the expense of their devoting that time to other students who would profit much more. Secondly, it raises the question of what mathematics can be learned by all, and whether such a minimum can justifiably be called 'mathematics'. Thirdly, there is growing evidence (e.g. from the Second International Mathematics Study, SIMS) that certain parts of the canonical mathematics curriculum are found to be very difficult by children everywhere. This raises considerable problems of motivation amongst those who do not succeed with the subject or who do not see the point of studying it, a fact that must be taken into account in syllabus construction.

There are broader issues, too, which lie outside the purview of mathematics educators but which are strongly influential in determining a country's policy on this issue. Assumptions held within a society about the distribution of intelligence within a population, or the possibility of generating intelligence, may be determining factors in this as in other aspects of educational policy.

In weighing up these and other factors, policy-makers must ask themselves which should be accorded greatest weight in the 1990s; answers from earlier decades may no longer be valid in the technologically new circumstances into which we are moving. To summarise this central issue:

Should mathematics be compulsory throughout a student's secondary education?

Alternative 1.   Yes.

Consequences: 1.   More mathematics teachers are needed, and many of these will spend time teaching to little apparent effect to students who have little interest and/or ability in the subject.

                2.   No student will be handicapped in either personal or career terms by being allowed to drop mathematics prematurely.

                3.   Social equity will be respected.

                4.   If only one form of mathematics is taught/ available to each age cohort, then the later stages of secondary schooling may well reveal serious problems of student motivation in mathematics.

                5.   Weaker students will have a sense of limited achievement or of failure reinforced right up to the time they leave school.

6.   Major issues of differentiation of curricula will have to be faced.

Alternative 2.   No.

Consequences: 1.   Mathematics teachers will spend their time teaching students who wish to be taught the subject.

2.   Motivational problems will be minimised.

3.   Mathematically weaker students will be able to spend their time on activities at which they can achieve greater success and satisfaction.

4.   Some students will be handicapped in later life.

5.   Very careful thought will need to be given to the optimal stage at which mathematics becomes optional, and to the criteria by which an individual student should be allowed to discontinue its study.

Whatever policy is adopted, education should clearly continue to have a substantial numerical-spatial-graphical component and this should be planned. To have value for all students this must be set in a context to which they can relate. It is an open question, and one to which the answer may not be the same for all students or in all countries, whether this component of education should take place in classes called 'mathematics', or in the context of studies related to design, to social studies, to a study of the physical world, and so on.

Certainly there is no place for compulsory mathematics taught as a set of rules and unexplained procedures. Education should be fundamentally rational, and in mathematics this implies that it should emphasise relationships between items of numerical and spatial knowledge. For example, the uses to which a particular geometrical shape can be put depends on the properties of that shape, and the various properties are not independent pieces of knowledge but are connected with each other. Again, in learning to handle number efficiently, it is as important to appreciate the connections between multiplication and division, and to recognise the situations in which they arise, as to be able to carry out the appropriate algorithms. This relational aspect of mathematics becomes increasingly significant as electronic devices become available to carry out routine procedures.

This does not imply that there is educational benefit for more than a small minority of school students in a form of mathematical education that emphasises definitions, axioms and a logically structured body of knowledge. Whilst many children can enjoy and appreciate some experience of exploring numbers and shapes as abstract entities, it is only the few for whom it is worthwhile to formalise this experience into "mathematics" as a formal discipline. In this sense mathematics should not be compulsory at any level.

6. The Unfinished Curriculum

The problem of school drop-outs has already been mentioned. In many countries of the Third World it is the major feature of the education system, particularly during the primary years, and it is likely to remain so for many years to come despite heroic efforts by governments to provide facilities for all their children to complete at least a primary education; indeed, there is a strong trend to extend the compulsory phase of schooling to a nine-year period of 'basic education'.

The drop-out problem faces the curriculum developer with a very difficult problem, the more so as drop-outs occur at every stage of the system. Should they just be ignored? Traditionally, the mathematics curriculum has been developed by a 'top-down' process (Committee of Inquiry, 1982). In this process, each stage is seen as a preparation for the next: primary prepares for lower secondary, lower secondary for upper secondary, which in its turn prepares for tertiary education. Yet by this stage only a minute proportion of the age cohort may remain within the formal education system; has all this 'preparation' been a waste of time and effort for the great mass of the students who are no longer there?

Even within countries where secondary education is effectively compulsory there is a problem of de facto drop-out, as lower achievers get irretrievably lost and simply switch off as far as mathematics is concerned. Have their years of study of the subject before that point been entirely wasted?

The issue can be seen in economic terms: how can one maximise the investment of time, of money and of human effort in the mathematics that has been taught to the dropping-out student? At present if a child drops out before the end of primary school – as the great majority still do in many countries of Africa, Asia and Latin America – then that child has gained practically nothing from the study of mathematics. Failure to complete a cycle of schooling results in nothing but a reinforced sense of failure. The ideal to aim at is to develop a curriculum such that a child who stays only one year in school gains one year's benefit; but is such an aim attainable?

How may the interests of students who drop out of school education prematurely be taken into account in planning the school mathematics curriculum?

Alternative 1.   Ignore the drop-outs.

Consequences: 1.   No changes would be needed: this is what curricula world-wide do at present.

2.   It is relatively easy to construct curricula
        which reach natural end-points only at the few
        official school-leaving stages.

3.   The curriculum caters well for those who
        proceed to the next stage of education.

4.   There is an enormous waste of human and
        material resources in those who drop out
        prematurely.

5.   Society will continue to consist largely of
        people whose attitude to mathematics is
        dominated by a sense of failure, and who find
        it impossible to learn more mathematics should
        it prove necessary in the future.

Alternative 2.   Aim to round-off the curriculum at each stage
                        where there is major drop-out.

Consequences: 1.   Curriculum construction becomes extremely
                        difficult.

                2.   There is potential mis-match between such a
                        curriculum, which would effectively preclude
                        propaedeutical work, and the cumulative nature
                        of an individual student's mathematical
                        knowledge.

                3.   The progress of those who do not drop out may
                        be slowed down.

                4.   It may require different curricula in different
                        parts of the same education system (e.g.
                        urban/rural), with a consequent shift to more
                        locally-based curriculum development, and
                        perhaps a socially unacceptable form of
                        differentiation of education and
                        opportunities.

                5.   This style of curriculum would most easily be
                        based more on student experience, and less on
                        mathematics itself, than at present.

                6.   Some people will question whether such a
                        curriculum would still be mathematics.

                7.   Those students who terminate prematurely will
                        gain the maximum possible value from however
                        many years of mathematics teaching they have
                        received.

Alternative 3.   A compromise strategy, in which each component
                        of the mathematics curriculum is seen as a
                        module which extends over a limited period of
                        not more than three years.

Consequences: 1.   Such a curriculum is harder to construct than
                              the traditional top-down variety.
                  2.   Applications of each mathematical topic
                              included must be sought, and taught, as quickly
                              as possible.
                  3.   There is a danger, also present in Alternative
                              2, that the curriculum will be mathematically
                              unbalanced.
                  4.   As a development of the 'spiral approach', this
                              strategy is less unfamiliar than alternative
                              2.
                  5.   Some, but not all, drop-outs are catered for.

        The optimum strategy in a particular education system will
depend largely on the termination profile within that system. There is
no perfect solution short of eliminating drop-out and mass absenteeism
from schools; this is beyond the capacity of mathematics educators to
achieve, being dependent on social and economic factors, though if
mathematics can be taught and learned in such a way as to give students
more experience of success it would go some way to help. In any case
we must beware of thinking that any education is 'complete'.

        While it may be impossible to eliminate the drop-out problem,
some countries are experimenting with the idea of allowing drop-outs to
drop back into the system later, perhaps by means of vouchers which
guarantee a specified number of years of full-time education and which
may be used after the normal school-learning age. Such a scheme has
major advantages of motivation at a period of life when people will be
aware of their real mathematical needs, in employment and in daily
life.

        Any attempt to cater for the problem of multi-stage drop-out
will result in a curriculum with greater emphasis on mathematical
literacy (a wider term than 'numeracy'), whereby the development of the
ability to think intelligently and critically in numerical and
quantitative terms will play a greater part, if necessary at the
expense of a certain amount of technique. Few students in their final
year of formal mathematics education require more technique; they are
better served by a curriculum that seeks to help them to appreciate the
role and power of mathematics in its applications at the level which
they have reached.

7. The Differentiated Curriculum

        In every education system the mathematical attainment of
students ranges widely, whether or not the subject is compulsory.
Recent research has suggested that in England there is a "seven-year
difference" in terms of what most 11 year old children might be
expected to be able to do. For example, most can answer correctly the
question "what number is 1 more than 6399?". There are some 14 year
olds who cannot do it, and some 7 year olds who can. Similar

comparisons can be made in respect of other topics (Hart, 1981). In
the light of such findings, it seems absurd to expect every child to
progress through the same mathematics curriculum at the same speed.
Whether such differential achievement stems from different inborn
ability, as teachers in many countries would say, or from sociological
and environmental factors, as would be thought in other countries, is
not our immediate concern. The simple observed fact is that
achievement varies, and varies dramatically, in school mathematics, and
this fact requires us to consider the need to offer different
mathematical diets to different students or groups of students. We
will use the word "differentiation" for such a policy. Its desirability or otherwise will be considered in alternatives-
and–consequences form in the next chapter. Here we consider the
various forms that it may take, and the strategies that can be used to
achieve it.

        In earlier decades, differentiation has often been by schools.
In some countries, independent schools have had – or have been thought
to have had – higher academic standards than government-funded schools,
and many have set entrance examinations which reinforce the difference.
Yet although such differentiation was ostensibly based on student
ability, in practice it was largely determined by social class and by
the ability of parents to pay the fees. In many countries some schools
are accepted as élite even within the state system of education;
although they may follow the same official curriculum and take the same
public examinations as other schools, in fact they proceed faster and
more thoroughly through the curriculum and their pupils achieve greater
examination success. Other systems allow for different kinds of
school, e.g. 'grammar', 'technical', 'general', within which different
mathematics curricula operate.

        Other strategies for differentiation may be used within a
school. The most common are 'setting' (whereby the students in a
particular school year are placed in groups for mathematics lessons
according to their previous levels of attainment in the subject), and
'streaming' (where students are grouped on criteria of more general
academic achievement, embracing other subjects as well as mathematics).
Where there are objections to such strategies, for example on
sociological grounds, the class teacher may adopt group work or
individualised learning materials in order to cater appropriately for
different students within a 'mixed-ability' class. Whatever system is
in force, sooner or later it results in different children doing
different things according to their own rate of progress in the
subject.

        At the level of the individual classroom, of course, the good
teacher already provides some curriculum differentiation; he will pose
more complex problems to the higher-achieving students in order to
stimulate them further, and less involved questions to the lower

attaining students in order to build up their confidence and create a
more positive attitude to learning mathematics. Our concern in this
study, however, is with policy issues.

Differentiation may itself take various forms. It may be based
on content, with weaker pupils following a watered-down version of that
studied by their more able contemporaries; an example is the two levels
of public examination at 16+ in England, the General Certificate of
Education (GCE) Ordinary Level for the most able quartile, and the
Certificate of Secondary Education (CSE) for the two middle quartiles
of the ability range. It is interesting to note that social pressures
are resulting in a national integration of these two examinations,
though in practice students of different abilities will still take
different papers, with still more marked differentiation of mathematical content, within the overall scheme.

Another form of differentiation is by speed of progress through
a common curriculum, with the obvious implication that the less able
the students, the less far will they have progressed through it by the
time they leave school. In some countries such differentiation is
effected by retaining students in a particular year until they attain
an acceptable degree of competence. In others all students are
promoted every year and differentiation is achieved through the use of
individualised learning materials, for example workcards, worksheets
and topic booklets. The advent of micros in the classroom greatly
extends the range of possibilities for such individualised curricula.

A variation is for all students in a class to progress through
the same set of topics at the same speed, but for the more able to
study the material at greater depth, to tackle harder examples and to
explore a greater range of applications. This is a time-honoured
technique with which most mathematics teachers are thoroughly familiar,
whatever the 'official' position on differentiation.

A more radical alternative is to construct independent
curricula for different ability groups of students. For example, the
balance between content and processes, discussed earlier in this
chapter, might be very different in a set of such curricula.

We shall look in more detail at the issue of differentiation in
the next chapter. However at this stage it is appropriate to add some
comments on the different stages of schooling. Very few state-funded
primary schools provide for differentiation except for some remedial
work for the weakest and some extension work for the most able. It is
crucial that at this stage low attainers must not become irretrievably
lost through being made to work through a syllabus too fast, or that
high attainers are condemned to doing scores of repetitive exercises
with the sole reward of collecting pages of ticks. The need is
therefore for remedial and extension material to supplement for the few
the standard material used by all. With the arrival of new technology
in the classroom, there is a danger that calculators, micros and other

exciting aspects of mathematics might be reserved for extension
activities with the most able. Such a policy would be disastrous to
the majority, and deprive them of the stimulation that such activities
can provide for all. Widespread availability of calculators and micros
can in fact help to make the classroom curriculum more flexible.

In those education systems within which differentiation is the
official policy, it usually starts in the middle grades of school
(Junior High School/lower secondary level). The various strategies
have already been described.

8. The Modular Curriculum

At the Senior High School/upper secondary level, a curriculum
constructed on a modular basis offers one means of differentiation. The
"core plus options" pattern is increasingly favoured, having many
advantages. Such a core-plus-options model is being proposed in Japan
for the final two years of secondary education (Fujita, 1985). The
core can be presented in different ways and at different depths. Some
options could be embodied in individualised material (not necessarily
in the form of a programmed text, but presented through a variety of
media) and would serve the more general purpose of helping the student
to learn how to learn.

Such a modular curriculum structure could mean a break with the
traditional one classroom/one teacher/one timetable. In the final year
or two of school, the pattern might rather be that of a mathematics
resources centre, with some timetabled classes, a variety of learning
resources available, and a tutorial staff on hand for consultation at
specified periods of the day.

9. The Cultural Curriculum

Reference was made in Chapter 2 to ethnomathematics, that body
of ideas and techniques which young children derive from their social
environment and with which they arrive at school. It was argued that
the teaching of mathematics, particularly in primary classes, should
seek to build on ethnomathematics rather than ignore it. More
generally, the curriculum planner must ask what is the specific
cultural significance of mathematics within any particular national
context. The geometry of Islamic art and architecture, and the proud
history of numbers and number theory in India, are among more obvious
examples. Reference in school work to the history of a nation's
mathematics can help to build up a sense of pride and of 'ownership' of
mathematics, reinforcing students' confidence and relating the subject
to history and to national tradition.

It must be remembered too that mathematics is the scientific
and intellectual field in which it is easiest for young people to
demonstrate precocious talent. The various national and international
olympiads are but one expression of this, and many examples can be

found in the history of mathematics. This can give rise to national pride, as in Poland and Hungary after 1920. On the other hand, the feeling that a nation is somehow lagging behind others in its overall mathematical attainments can lead to what appear to be over-dramatic reactions, for example the massive post-Sputnik curriculum upheaval in the USA in the 1950s which led to the 'new math' era, and more recently to that same country's series of reports which led to proposals to ensure that American students' mathematical "achievement is the best in the world by 1995" (National Science Board, 1983).

10. Summary

        This chapter has surveyed some major general issues concerning the future of school mathematics: its status, its world-wide similarities, the difficulties of relating it effectively to other aspects of a student's education and experience, the balance between content and process, the extent to which it should be compulsory for all, the problems of differentiation and of drop-out, how it may both reflect and contribute to the particular culture in which it is taking place.

        It is to be hoped that those responsible for the 1990s mathematics curriculum in national education systems will seek to get to grips with such issues, and not be content merely to tinker with the more superficial details of whether this topic or that should be included in the syllabus for a particular school year. Only by this more fundamental reappraisal will school mathematics, country by country, more effectively serve the future needs of citizens into the next millenium.

Chapter 4

The Content of the School Mathematics Curriculum

1. The Present Position

        We have referred earlier to the remarkable degree of uniformity
of content so far as mathematics courses for the general secondary
school pupil is concerned. Before beginning to look ahead, then, it is
useful to see in which areas of the curriculum there are generally
accepted goals. Here we shall draw on the findings of the Second
International Mathematics Study (SIMS) and to its investigation of the
courses followed by 'all students in the grade in which the modal
number has attained the age of 13.0 - 13.11 years in the middle of the
school year' (see Travers et al, in press).

        First it is essential to draw attention to the three levels on
which the content of the school mathematics curriculum can be viewed:

    (a)   the intended curriculum: what is prescribed in national
             and examination syllabuses;

    (b)   the implemented curriculum: what teachers teach;

    (c)   the attained curriculum: what students learn.

        The easiest 'curriculum' to investigate is the intended, for it
is this which is printed in official syllabuses.

        Below we give a list of topics which some twenty countries
(developed and developing) were asked to comment upon and to assign to
them the weights very important (V), important (I), or not important
(-). The composite weightings are shown in Table 4.1. (Here Is
indicates important in some countries.) The cognitive level of
behaviour expected from students on the topics is also shown, thus for
example, a deeper grasp of decimals is usually required than of square
roots.

Content Topics	Behavioural Categories			
	Computation	Comprehension	Application	Analysis
Arithmetic				
Natural numbers and whole numbers	V	V	V	I
Common fractions	V	V	I	I
Decimal fractions	V	V	V	I
Ratio, proportion, percentage	V	V	I	I
Number theory	I	I	–	–
Powers and exponents	I	I	–	–
Other numeration systems	–	–	–	–
Square roots	I	I	–	–
Dimensional analysis	I	I	–	–
Algebra				
Integers	V	V	I	I
Rationals	I	I	I	I
Integer exponents	Is	–	–	–
Formulas and algebraic expressions	I	I	I	I
Polynomials and rational expressions	I	Is	–	–
Equations and inequations (linear only)	V	I	I	Is
Relations and functions	I	I	I	–
Systems of linear equations	–	–	–	–
Finite systems	–	–	–	–
Finite sets	I	I	I	–
Flowcharts and programming	–	–	–	–
Real numbers	–	–	–	–
Geometry				
Classification of plane figures	I	V	I	Is
Properties of plane figures	I	V	I	I
Congruence of plane figures	I	I	I	Is
Similarity of plane figures	I	I	I	Is
Geometric constructions	Is	Is	Is	–
Pythagorean triangles	Is	Is	Is	–
Coordinates	I	I	I	Is
Simple deductions	Is	I	I	I
Informal transformations in geometry	I	I	I	–
Relationships between lines and planes in space	–	–	–	–
Solids (symmetry properties)	Is	Is	Is	–
Spatial visualization and representation	–	Is	Is	–
Orientation (spatial)	–	Is	–	–
Decomposition of figures	–	–	–	–
Transformational geometry	Is	Is	Is	–

Descriptive Statistics

Data collection	Is	I	I	–
Organization of data	I	I	I	Is
Representation of data	I	I	I	Is
Interpretation of data (mean, median, mode)	I	I	I	–
Combinatorics	–	–	–	–
Outcomes, sample spaces and events	Is	–	–	–
Counting of sets, $P(A \cup B)$, $P(A \cap B)$, independent events	–	–	–	–
Mutually exclusive events	–	–	–	–
Complementary events	–	–	–	–

Measurement

Standard units of measure	V	V	V	–
Estimation	I	I	I	–
Approximation	I	I	I	–
Determination of measures: areas, volumes, etc	V	V	I	I

Table 4.1.   Content coverage at 13+ as revealed by SIMS.

With these general findings in mind, appropriate test items were constructed and national representatives within participating countries were asked whether or not these items were appropriate within their system.

On average, the percentages of questions found appropriate were:

arithmetic	92
measurement	91
algebra	83
statistics	69
geometry	64.

There was, therefore, a considerable degree of uniformity so far as national intentions were concerned about the teaching of arithmetic, measurement and algebra. Vastly differing emphases were placed on the teaching of statistics, but there was general agreement on what should be taught in those countries which taught it. There was relatively little agreement on where emphases should lie in the teaching of geometry.

These, however, were national intentions. What was revealed when teachers were asked whether or not they had given their students the necessary background knowledge to attempt individual questions? The items on measurement had been almost uniformly well covered, as had those on arithmetic (with the exception of square roots). Algebra had been slightly less well covered than arithmetic with the exception of

exponents and indices. That is, those parts of mathematics on which
countries had all placed emphasis were in general stressed by teachers.
On the other hand, statistics and geometry tended to be ignored by
teachers even where they were on the syllabus. Some geometrical topics,
such as coordinates and plane figures, were given the intended
emphasis, while others, such as constructions and Pythagoras, were not.

There would seem to exist, then, for teachers at any rate, a
basic 'core' curriculum consisting of arithmetic, algebra and
measurement, to which other content is added partly on the whim of the
nation, partly on that of the teacher.

What must be emphasised here, moreover, is that this degree of
common practice at age 13+ to be observed within those countries
participating in SIMS is achieved despite:

(a)   differences of up to two years in the beginning of
        compulsory schooling;
(b)   that, because of the different ages of transfer from
        primary to secondary school, some students were in the
        last year of primary school and others in the third
        year of secondary;
(c)   enormous differences in the (national mean) percentage
        of the total timetable allotted to the teaching of
        mathematics (9% to 17%) (within individual countries
        the S.D. of hours devoted to mathematics within
        individual schools varied from 0 in a highly
        centralised system to as much as a third of the
        national mean number of hours).

Perhaps not surprisingly the mathematical content which
teachers would seem actually to teach corresponds to some of the
'basic' or 'foundation lists' to be found in recent reports (e.g.
National Science Board, 1983, Committee of Inquiry, 1982). The list
given in the former, pp. 93-94, for instruction at the K-8 level (i.e.
ages 5-14) is:

"- Understanding of arithmetic operations and knowledge of when
   and where specific operations should be used.

 - Development of a thorough understanding of and facility with
   one digit number facts.

 - Ability to use selectively calculators and computers to help
   develop concepts and to do many of the tedious computations
   that previously had to be done by using paper and pencil.

 - Development of skill in the use of informal mental
   arithmetic, first in providing exact answers to simple
   problems and later, approximate answers to more complicated
   problems.

- Development of skills in estimation and approximation.

- Development of problem-solving abilities. Trial and error
  methods, guessing and guestimating in solving word problems
  should be actively encouraged at all levels.

- Understanding of elementary data analysis, elementary
  statistics, and probability.

- Knowledge of place value, decimals, percent, and scientific
  notation.

- Understanding of relationship of numbers to geometry.

- Understanding of fractions as numbers, comparison of
  fractions, and conversion to decimals.

- Development of an intuitive geometric understanding and
  ability to use the mensuration formulas for two- and
  three-dimensional figures.

- Ability to use the concepts of sets and some of the language
  of sets where appropriate. However, sets and set language
  are useful tools, not end goals, and it is inappropriate to
  start every year's program with a chapter on sets.

- Understanding of elementary function concepts including
  dynamic models of increasing or decreasing phenomena.

- Ability to use some algebraic symbolism and techniques,
  particularly in grades 7-8."

The 'Cockcroft' Committee of Inquiry Report has no formal
algebra or set language in its basic 'foundation' list (intended for
all students in the 11-16 range: the middle and high-attainers would
be expected to cover more) and restricts geometry or, as it describes
it, 'spatial concepts', almost entirely to drawing and describing the
properties of common plane and solid shapes, and to map reading. Like
its US counterpart, it gives considerable emphasis to statistical
ideas, graphs and pictorial representation, and to the use of the
calculator.

The picture that one obtains from the SIMS findings and which
is echoed in other national reports is that many children currently
meet only a very basic form of mathematics, comprising mainly
arithmetic and mensuration, and oriented towards utilitarian ends, and
that even when more is envisaged by national curriculum designers, it
is not always implemented in the classroom.

The position in the later years of schooling is very different. One extremely important difference arises from the fact that although at 13+ mathematics is taught to all, the percentage of 17 year-olds still in school varies very much from country to country (from less than 20 to over 90 per cent in those studied by SIMS), whilst of those still at school differing percentages are following a 'substantial' mathematics course (from 10% to 100%). Thus the actual percentage of 17 year-olds following a pre-university mathematics course (i.e. a course which would allow them to study mathematics at university) varied within the SIMS countries from 6 to 50.

When one takes into account differences in the number of timetabled hours allotted to mathematics, which varied for individual students from less than 10% of the school week to almost half, one would expect to find very different national patterns. Nevertheless, there is still a minor degree of uniformity, particularly in basic algebra, sets and relations, elementary functions and number systems. There is a great variety amongst countries in geometry content, and few countries at present include significant amounts of statistics and probability in their syllabuses. There is a sharp contrast in the treatment of the single-variable calculus, ranging from nil to a very full one.

On the other hand, at this level the implemented curriculum would seem more closely to follow the intended one than at 13+.

By mid-1986 only a minority of the countries participating in SIMS had published their national reports concerning student attainment at 13+ and 17+. It is hoped, however, that international reviews will be published in 1987 (Garden and Robitaille). These should assist identification of those topics in the curriculum which are universally found easy/difficult, and, perhaps more interestingly and of more value for those planning changes, those items which seem to cause difficulty to students in some countries but not in others.

We have referred to the necessity of always keeping in mind the three curriculum levels – intended, implemented and attained. This is especially important because when curricular issues are discussed in reports such as this present one, it is inevitably the intended curriculum which claims most attention. The transition from intended to implemented curriculum is a vital part of curriculum development dependent upon the cooperation and, often, re-education of teachers. It is more the concern then of later chapters than of this. Yet syllabus design cannot be successfully undertaken without due consideration being given to ensuring that great discrepancies between the intended and the implemented curricula will not arise, and, of course, to ensuring that what is taught is closely linked to what can be learned!

2. Arriving at the present position

Although the preceding section might give the (correct?) impression that in general school mathematics is impoverished and in the main directed towards utilitarian ends, it must not be forgotten that it is an impression gained by considering what is being taught to the majority of students and that the SIMS items were designed to be generally acceptable rather than to test some of the more individual aspects of nations' curricula.

Certainly, it was not the case that the reformers of the 1960s sought such goals, neither is such a limited curriculum the norm in many types of school.

Until recent years schools could be roughly classified into two types (in some countries this classification is still valid): the elementary/vocational school and that intended to prepare students for university entrance. The traditional curricula, i.e. prior to the 1960s' reforms, are summarised in Figures 4.1 and 4.2

We see that mathematics for the majority was confined to arithmetic and mensuration, whereas those few who went to a grammar school/gymnasium/lycée were offered a much more widely based course which prepared them for university mathematics. (Indeed, to a large extent the course failed to make educational sense unless students did eventually go to university.) The course as described in Figure 4.2 evolved during the latter part of the nineteenth century and changed little until the 1960s. (We note that in North America the curriculum was 'banded' more by grades, e.g. Grade 10 was devoted to geometry, and did not traditionally include the calculus.)

The reforms of the 1960s took a variety of forms, but in general, their aims can be summarised as in Figure 4.3. Here we note that the mathematics of the primary school has been completely rethought, that 'structures' (primarily algebraic) now underpin the secondary school curriculum (the extent to which these were made explicit varied from country to country), and that, in many countries, geometry as a separate entity tended to disappear from the curriculum although, perhaps in compensation, increased use was made of pictorial and graphical representation. In particular, coordinate geometry tended to move down the school. (Unsuccessful attempts were made in the USA to break away from the old 'banded' curriculum and to replace it by the 'spiral' curriculum favoured in Europe; attempts which demonstrated yet again the great difficulty of making structural changes within the school curriculum.)

Fig. 4.1 Learning track of arithmetic

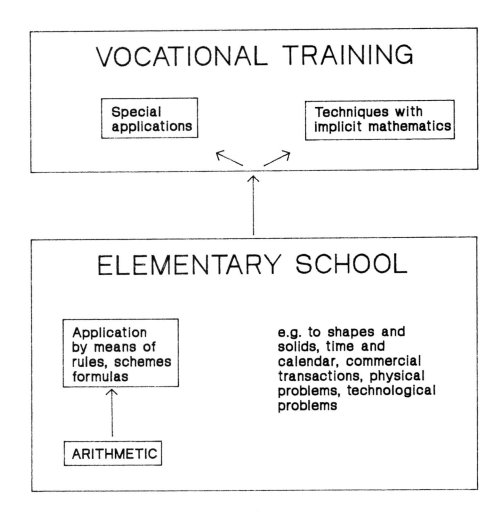

Fig. 4.2 Learning track of traditional Pure Mathematics

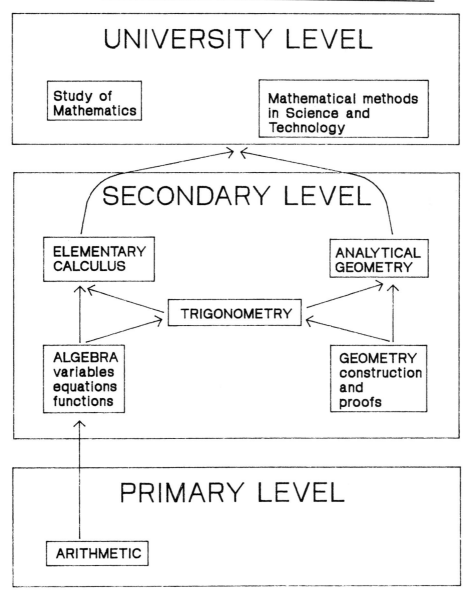

Fig. 4.3 Learning track of 'Modern' Mathematics

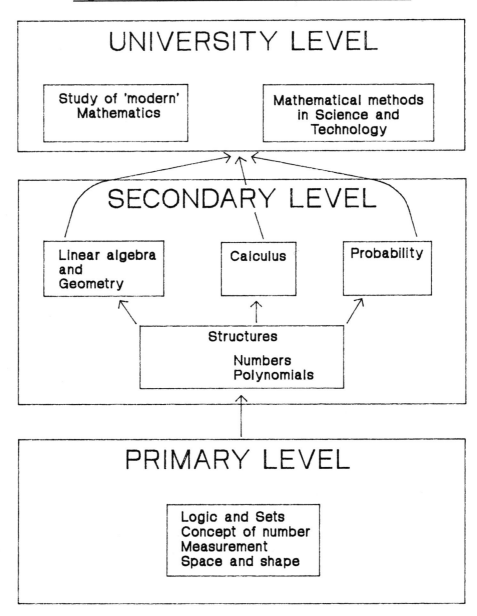

There has recently been a retreat from the positions taken up in the 1960s. For example, within the socialist countries of Eastern Europe there appears to be a common trend to remove emphasis from the structural and deductive aspects of mathematics: from a set-theoretical approach, algebraic structures, and axiomatic approaches. In some countries which introduced an axiomatic approach to geometry, recent trends have not just restored the status-quo, they have rejected even the 'loosely axiomatic' approach favoured pre-1960. Thus, for example, for the first time in the history of mathematics education in Poland, geometry is now being taught without any axiomatic or pseudo-axiomatic base. Similar remarks could be made concerning curricula in other countries. (These trends are, of course, closely connected with problems relating to 'mathematics for all' and policies relating to differentiation or the lack of it.)

As was mentioned above, not all the innovations have been abandoned. We have already mentioned the earlier introduction of coordinate geometry (although there are now fewer attempts to develop it algebraically at an early stage). Probability and vectors (but not vector spaces) seem to have become established in the curricula in many countries, and equations as a tool for solving word problems are now introduced earlier.

These examples, though, only serve to emphasise that it has proved easier to introduce new techniques that can readily be put to use on collections of exercises, than abstract concepts: a fact that should cause no surprise, but which should not be overlooked when planning for the 1990s.

We also note attempts (see, for example, Committee of Inquiry, 1982) to cut down on the amount of mathematics taught. (It was a feature of the 1960s reforms that not only was content changed, but it was usually increased in extent, the argument being that with increased understanding of the underlying structures and less time spent on boring, repetitive exercises, more 'real' mathematics could be learned.)

Yet, because the level of mathematics taught in schools is so often viewed as a measure of a nation's technological and scientific strength (see Wirszup (1981) for an example of an application of this type of reasoning), there is still a great temptation to present a macho-mathematical image to the world. Thus, for example, until recently, by the end of the 8th form of their 10-year compulsory school course, (15-year old) students in the USSR were expected:

(i)   to have mastered all the basic school algebra up to quadratic equations, inequalities and systems of equations,

(ii)   to know and use linear and quadratic functions and their graphs,

(iii)   to know the basic properties of arithmetic and geometric
          progressions,

(iv)   to know and use rational exponents and powers,

(v)   to know the basic properties of logarithms, the graph of the
          logarithmic function; to be able to solve simple logarithmic
          and exponential equations and inequalities and (still, in the
          1980s) to be able to use logarithmic tables and the slide
          rule,

(vi)   to know all the basic geometry and trigonometry of the plane
          (there is none in the curricula of succeeding years), some
          approaches to geometry still being axiomatic.

          (In the following year all pupils began the study of the
          calculus.)

        The contrast with the restricted nature of the Cockcroft
Committee's 'foundation' list is truly amazing, and the syllabus
certainly helps to explain the impact in some quarters of Wirszup's
paper. The lowering (to 6) of the age of entry to schools in the USSR
has, however, resulted in a re-thinking of the whole school curriculum.
The general school course is now of 11 years duration and yet certain
topics (e.g. (iv) and (v) above) have been omitted from the curricula
of 15 year-olds. Again, we see evidence of moves to reduce syllabus
content.

        The two approaches, USSR and Cockcroft, can , however, still be
taken as typifying two approaches to curriculum planning and to the
problem of differentiation of students.

Alternative 1.   There is no differentiation and the curriculum is geared
to what the better students can achieve.

Consequences: 1. The better students reach a high standard.
                  2. The social consequences of differentiation and the
                     necessity for taking decisions about who does what
                     are avoided.
                  3. Because expectations are set high, then achievements
                     will usually be greater.
                  4. All students will be seen to have had an 'equal'
                     chance; failure cannot be ascribed to a lack of
                     opportunity.
                  5. Many students will flounder, understanding little
                     and becoming increasingly disenchanted with
                     mathematics and schooling. Differentiation will be
                     by whether or not the work is understood.
                     Discipline and other problems may arise.

6. Greater demands are put on teachers of unstreamed classes.

Alternative 2. Traditional differentiation (by school/streaming/ setting/individualisation) is practised, although it may be delayed as long as possible.

Consequences:

1. The teacher has a relatively homogeneous task.
2. The state can manipulate the labour force to meet requirements.
3. Individual students follow a curriculum suited to them, thus enhancing their prospects of success.
4. It permits differentiation of teachers, by qualification, salary and status.
5. Low achievers are labelled as such, with the reinforcement of self-fulfilling prophecy. It is not easy to cater for late developers.
6. Selection is difficult to effect and is often influenced by social class.
7. It sets up tensions between schools and students, and between different sections of education.
8. Schools help to consolidate social status, and are seen as much as institutions to preserve social structures as to provide education.

Alternative 3. Differentiation is by future occupational need.

Consequences:

1. Such a criterion would appear to be widely acceptable, but raises difficulties relating to who determines what the future occupation will be.
2. There would not necessarily be a match between a student's mathematical capacity and the curriculum he followed, except insofar as there is some correlation between future occupation and general intellectual capability.
3. Differentiation would have to be postponed until the late stages of a school career, when likely future occupation may be known. This will almost certainly be too late to prevent weaker students getting lost and more able students bored.

The problems of differentiation of students and of determining the 'level' at which a curriculum should be pitched are not easy ones to resolve (we have, for instance, referred above to the USSR decision to reduce the 'level' of part of its mathematical curriculum). Here a mathematical model can be usefully employed: that which sets out to determine the 'yield' from an educational system and the nature of that yield. We can, in a rough way, attach a 'weight' to any curriculum for its content in terms of both product and process. A general measure of an educational system's yield can then be obtained by finding the

weighted sum in which the weights of the various curricula are
multiplied by the proportion of students effectively following them.
Thus, for example, although the 'weight' of the USSR syllabus has been
reduced, if this results in more students coping with the course
effectively, the 'yield' will rise. There may, of course, be political
reasons why a country may wish to cultivate a high-achieving élite,
but, in general, the notion of 'yield' can be helpful when analysing
an educational system.

3.   On structuring the mathematics curriculum

        As we have indicated above, a major aim of the reforms of the
1960s was to strengthen the link between 'school' and 'higher'
mathematics. It was hoped that a properly 'structured' school course
would not only prepare students better for higher mathematics, but
would also prove more accessible, and could be learned by a greater
proportion of students, than traditional mathematics. A major premise
was that by gradually revealing to pupils the structure of mathematics,
as it was perceived by twentieth century mathematicians, they, the
students, would more readily comprehend, use and appreciate
mathematics. It did not work out quite so simply, and, as we have
noted, reaction set in.

        The movement to base a unified treatment of mathematics on
algebraic structure (for topological structures were never widely
considered at a school level – although, for example, Papy and, for a
very brief period, SMP made essays into this field) effectively
supplanted another attempt to 'structure' mathematics education by
dividing mathematics into the topic areas of arithmetic, algebra and
geometry. Whilst such a partitioning might have been mathematically
undesirable, it did offer certain benefits for the student, for each
branch was clearly associated with certain patterns of thought and
action. It was in geometry one went 'To prove, given,
construction,...', and in algebra one made use of lower case letters
and equations. The support for learning may have been naive and dearly
bought, but nevertheless it existed. The structural approach of the
1960s, although often accompanied by pedagogical arguments, failed to be
as closely attuned with learning patterns as had been hoped and
claimed. Yet did it deserve the fate it has received? Were its main
faults, not its insistence on providing the student with clear
mathematical structures, but rather the degree of complexity of the
structures which were presented and the excessive speed of
presentation?

        The answer to this last question would appear to be 'Yes', but
then we are faced with the more difficult question: 'How is the
learning of mathematics to be structured, and how is the students'
acquisition of such structures to be facilitated?'

Valuable theories and examples on the creation of 'didactical structures' are to be found in the writings of Freudenthal (in particular, 1983), and no one seriously considering the construction of mathematics curricula for the 1990s should fail to become acquainted with these. For example, Freudenthal makes important criticisms of the way in which the structures presented to students in the 1960s were based on mathematicians' hindsight, rather than on students' (and mathematicians') learning patterns: they were examples of 'didactical inversion'. It is impossible to summarise Freudenthal's theories here, we can only refer readers to his works which represent one person's attempt to begin to answer the key question:

How can we structure the mathematics curriculum so that the underlying framework not only facilitates learning, but also illuminates our mathematical objectives?

This is in many ways the key question of curriculum design. Yet even it must yield in order of priority to:

What is it that we want students to learn?

As we have already indicated, when we attempt to answer this last question we have to take account of both 'products' and 'processes'. We shall now try to provide additional guidance by briefly considering various types of learning and things to be learned. First we observe how when 'mathematical knowledge' is mentioned the suggestion so often is of acquiring 'acknowledged' mathematics, the mathematics of the past, the mathematics that specialists (and some laymen) recognise. Yet it is useful to make the distinction between 'knowing how' and 'knowing that', a distinction which reminds us of the process/product dichotomy. Indeed, English suffers in that 'knowledge' has no plural and for that reason we are apt to neglect the possibility of distinguishing between different types of knowledge. Thus, for example, the coming of information systems which we mentioned earlier emphasises the need to learn 'knowledge where'.

Mathematical knowledge exists, too, on a variety of levels: one can 'know' Pythagoras' Theorem (as a statement), one can 'know' how to use it to solve particular types of problem, one can 'know' how to justify it (and it is possible to 'know' how to do this in either only one or in a variety of ways: the former suffices for one to 'know' that it is true —under certain usually unstated assumptions - the latter clearly increases in some way our 'knowledge'), one can 'know' how it can be generalised, one can 'know' how the result related to other items of knowledge (e.g. the circular trigonometric functions), one can 'know' that there is a theorem due to Pythagoras which says something about triangles and areas and 'know' where to turn for further details,.... (Such levels of knowledge have been enunciated on many occasions - for example, several centuries ago by Spinoza.) Above these types of knowledge is a form of meta-knowledge: 'knowing' how to make use of these lower levels, i.e. knowledge for action.

How might our curricula cater for these different aspects of knowledge?

A quote from Freudenthal (1977) sums up, not only the complexity of mathematical knowledge and understanding, but the crucial difference between the understanding which teachers must seek and that which we wish learners to gain.

'There are so many kinds of understanding in mathematics. At every moment you may believe that you have just reached ultimate understanding of some subject, such that nothing is left to be desired. But no, there is no ultimate understanding in mathematics, you can understand any problem in an ever larger context, from an ever higher point of view; and finally – it looks the lowest of all, but perhaps it is the highest – you can learn to understand it in the perspective of the learning child".

This difference of viewpoints can also be seen to exist in another important way. In our discussion so far, we have considered various types of knowledge which we, as teachers, wish our students to acquire. But in doing this will our aims be sufficiently explicit? What view of mathematical knowledge will our students absorb?

Most students will inevitably construct a type of knowledge, 'desirable' knowledge, and in the vast majority of cases it will be the knowledge which is required to obtain external qualifications. 'Desirable' knowledge tends to become synonymous with 'assessable' knowledge. There is a need then to ensure that evaluation and assessment procedures are developed to extend the range of 'assessable' knowledge so as to cover all the various types of knowledge.

Students, parents and employers will all form their own opinions on what constitutes mathematical knowledge. Often this will be restricted to a knowledge of facts and a few well-defined skills. This conception is likely to be reinforced by the media (particularly television). One effect of this is that it colours the student's view of knowledge and of how it is acquired. Certainly, mathematics is likely to suffer because of this. Mathematics as presented in the media is almost inevitably a presentation of acquired 'knowledge that', with the emphasis on the accumulation of information – gurus who 'know' about cosmology, topology,... Unfortunately, TV programmes showing Atiyah, Thom or other mathematicians actually doing mathematics and exercising those powers of creativity which bring them their distinction would hardly be feasible.

We are faced then with another key question:

How can students, educators and others be brought to a better understanding of mathematical knowledge in its various forms?

Very frequently, for the purpose of specifying curricular content and goals to others, desirable skills and concepts are listed in syllabuses, national criteria, etc. Describing learning, and prescribing teaching, in this way allows itemising, with the consequent danger that knowledge may thereby be atomised and, in a sense, finalised. However, whereas skills may be learned once-for-all (needing only to be practised or rehearsed) concepts are only locally stable and are subject to change.

Moreover, skills and concepts seem to be essential components of something larger. But of what?

Again, lists of goals for mathematics learning are likely to contain many 'how-to's:

How to solve problems, how to construct proofs, how to model situations, how to generalise results, how to identify crucial examples, how to search for and recognise counterexamples, etc.

Perhaps if these were analysed we could find constituent 'knowings', 'skills' and 'concepts'. Essentially, though, they are synthetic procedures with a significant 'managerial' component. Indeed, what is the use of any mathematical knowledge without the skill to manage it?

The development of 'managerial' skills is a key, but neglected, component at all levels. Although its lack is more keenly felt within higher education (at which stage students have more knowledge available to be managed), it still demands emphasis at a school level.

Knowing what one knows, how 'this' connects with 'that', how to deal with cognitive conflict ... is perhaps the most neglected aspect of all our learning. It cannot be written down in the form of behaviourally specific objectives and no pedagogical advice is available. Yet, again, it is an essential part of learning mathematics.

The points that we have raised above illustrate the range of questions which must be answered – if only in a temporary fashion – before we can begin to plan a new curriculum embracing both content and method. Clearly, more than a 'knowledge' of mathematical structure as set out by Bourbaki is required if the work is to be done successfully.

Yet it must not be forgotten that the 'structural' movement of the 1960s had appeal because it offered a framework within which the 'whole' of the curriculum could be viewed. We have here put 'whole' in inverted commas, because, of course, many educators felt that the curriculum they desired could not be emcompassed within the Bourbaki framework; that, for example, it provided no obvious role for either applications or learning how to apply mathematics.

Recently, there has been a reaction to such macrostructures. Indeed, following the work of Papert there is much talk of 'microworlds', incubators in which certain kinds of mathematical thinking can hatch and grow with particular ease (Papert, 1980). Such ideas are indeed rich and demand consideration by anyone planning a curriculum for the 1990s. However, it must be recalled that many problems of traditional mathematics learning would appear to have been caused by the establishment of 'microworlds' (in a somewhat different sense!) - for example, the formal microworlds of arithmetic and of differentiation techniques. These, of course, do not have the richness of a 'Papert' microworld, but they illustrate the learning and teaching problems associated with linking microworlds, applying knowledge gained in one to another, and creating meta-structures. It must also be remembered that one of the chief aims of a mathematician or scientist is to search for structures which unite different 'microworlds'. (For example, potential theory is the common structure of several worlds of mathematical physics, such as electrostatics, ideal fluids, etc.)

The computer will, as Papert has demonstrated, help us to provide many valuable 'microworlds' which can guide and assist learning. Ethnomathematics will furnish other examples. The problem facing the curriculum designer and teacher, then, will be to plan in order to coordinate, reinforce, and unite such learning experiences and to embed them within a growing mathematical framework which can be perceived, appreciated and utilised by the learner.

At the moment, our understanding of the problems involved is such that we can only have limited faith in the curricular structures and solutions which we propose. Our hope is that research and critical evaluation of existing work will provide future curriculum planners with a surer basis on which to make such decisions.

Chapter 5.

On Particular Content Issues

Any international seminar of the late 1950s or early 1960s would almost certainly have thought about desired curricular innovation in terms of content change: à bas Euclid; away with triangles; sets, relations, functions, linear algebra, and algebraic structure for all, ...

It is, then, perhaps of the greatest significance that at the Kuwait meeting, little, if anything, in the way of new mathematical content was urged (although there was a wish to see certain content, in particular probability and statistics, taught in all countries). Indeed, it was commonly felt that some present content would have to be sacrificed in an attempt to raise the general level of students' understanding and to foster the growth of other types of knowledge than those associated with rote learning.

Changes in the curriculum were viewed, then, more in the form of possible restructurings to reflect better-defined aims and students' learning patterns, than in introducing new content.

Such restructurings will not, however, be quickly accepted. Indeed, much exploratory work would seem to be required before large-scale recommendations can be made. Here it is interesting to note such experiments as that in Brazil where the teacher is expected to allocate 60% of classroom time to covering content which has been centrally stipulated, and 40% to activities which he or she selects from a growing à la carte menu. (The separation of these two strands of mathematics learning and teaching might be deprecated, yet perhaps the analogy of a cookery recipe in which the two constituents are gradually made to mix is helpful.)

Again, returning to the complex problem of differentiation, it would seem that the planning of differentiated curricula has been dominated by a 'content' view of mathematics based on a linear model which sees 'high-level' mathematics for the university bound at one end of the spectrum and 'low-level', utilitarian mathematics for citizenship at the other. Thus many countries formed their mathematics course for all by taking the traditional 'high-level' course and offering it to lower and middle-ability children in a diluted form. This approach was deprecated by the Cockcroft Committee in England

(Committee of Inquiry, 1982), but it could be argued that their
'bottom-up' approach essentially accepts the same linear model. In
their case the starting-off point is the 'low-level', foundation list
and topics are added to this with the intention that the high attainers
will reach 'high-level' mathematics, i.e. the various curricula are
'nested' with respect to content.

It is not obvious that this linear model is the one most
appropriate for 'mathematics for all'. Yet one suspects that in many
countries it will still be the dominant model in the 1990s.

Indeed, perhaps it was the hold that this model has in the
minds of mathematics educators which led to the lack of emphasis on
content in Kuwait. For the content of 'high' and 'low' level school
mathematics seems so well-established (subject to borderline arguments
concerning, for example, geometry), that, at the present time,
attention tends to be focussed on non-content curricular issues. Yet
the extraordinary developments in the applications of mathematics
brought about by the computer revolution may well offer a rich source
for experiments and developments with new content within a new model.
The dominance of the traditional 'linear' model for school mathematics
should, therefore, be challenged and alternatives explored.

Yet certain topics still demanded considerable attention and we
shall now look briefly at these.

1. Probability and statistics

Although probability and statistics tend always to be
associated with each other in discussions on curricula, their claims
for a place in the school curriculum differ, as do the problems of
teaching them.

First, however, it must be emphasised that both have an
overwhelming claim for inclusion in the school curriculum. As we have
indicated earlier, a knowledge of both is essential for good
citizenship today. As the research of Fischbein and others has shown,
the innate sense of probability is usually far too naive and will soon
lead to quantitative misjudgements. There is a need for that sense to
be developed and strengthened through mathematics education. The case
for statistics has been argued in Chapter 2 .

Fujita (1985) has argued that the purpose of mathematics
education is to cultivate 'mathematical intelligence', through a
combination of 'mathematical literacy' and 'ability for mathematical
thought'. He sees the latter ability as being less accessible than the
former and, as such, more appropriate, at least in its more complex
forms, for those who are going to specialise in mathematics: it
subsumes such abilities as deducing, proving, abstracting. On the
other hand, the goal of 'mathematical literacy' is one which should be
sought for all. In diagrammatic form, then, one obtains Figure 5.1.

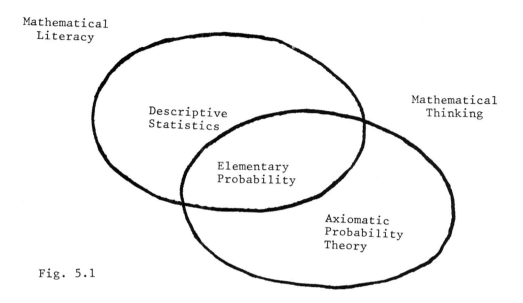

Mathematical
Literacy

Mathematical
Thinking

Descriptive
Statistics

Elementary
Probability

Axiomatic
Probability
Theory

Fig. 5.1

        Within this primitive model (which knowingly oversimplifies and
which, we note, presupposes a dichotomy between goals for all and goals
for some), we see that 'descriptive statistics' and 'elementary
probability' occupy different positions. The former lies clearly in
the domain of 'mathematical literacy'. At school level it is an
elementary introduction to the analysis and presentation of data –
something which becomes ever more important in a world in which the
amount of data which is stored, available and presented is increasing
rapidly. As such it provides an excellent vehicle for making links
with other subjects, such as Geography, and for practising graphical
and arithmetical skills (including those acquired through use of a
calculator). It cannot, of course, be done without 'thought', but it
is not in the syllabus even secondarily as an agent for fostering
mathematical thought, and many would argue that, except for a small
minority of students, it is inappropriate to pursue statistics at a
school level beyond this first, descriptive stage. (Some advanced
students may well require certain statistical techniques as tools, but
providing justification for these within school mathematics lessons is
not easily accomplished.)

        On the other hand, once one has progressed beyond merely
observing and recording the results of die-throwing, coin-tossing,
wheel-spinning, the generation of random events with a micro, or
whatever other apparatus is employed, probability becomes inextricably
linked with patterns of mathematical thought. These can, of course, be

encountered on many levels and through a variety of different examples. In general, though, the teaching of probability would seem to offer far more opportunities at a school level for fostering mathematical thinking than does statistics.

2. Geometry

No particular mathematical area within the school curriculum arouses so much concern amongst mathematicians as does geometry, the teaching of which has undergone a total transformation in the last thirty years or so. That concern is echoed by many mathematics educators (see, for example, Belgian ICMI, 1982) and a considerable, but declining, number of teachers (for many teachers nowadays do not have a 'Euclidean' past to recall).

We have already drawn attention to the way that a formal treatment of geometry has vanished from many school systems. In others, formal treatments remain, but few, if any, give indications that they are more successful - either in being accessible to students, motivating them, or preparing them better for future mathematical work - than the Euclid which they replaced.

What has happened in the last three decades is that, free from their traditional shackles, teachers and educators have developed a rich variety of geometrical examples and illustrations which can lead pupils to an excellent understanding of space and of elementary geometrical figures (see, for example, the published works of IOWO in The Netherlands and the School Mathematics Project in England). Moreover, the introduction of LOGO and other software into classrooms has created new possibilities for student experimentation, particularly within geometry. Perhaps even more challenging, computer-based opportunities for transforming geometry teaching in the 1990s will be provided by computer assisted design software which at the moment has had little impact on schools.

The result of such development work is that the spatial aspects of mathematical literacy (to use Fujita's model again) can be well catered for within the school curriculum. Yet the problem of making the transition from activities designed to provide a 'feeling' for space, and which can be seen from a philosophical standpoint as part of physics rather than mathematics, to 'geometrical thought' in the traditional sense has not been successfully solved. For the majority of students this transition may not be possible, and it could be argued that in an overall educational context this creates no great problem. As a result of recent curriculum changes, the 'few', the mathematically-gifted students, have at their disposal powerful algebraic tools, coordinates and vectors, which they can apply to geometric problems. They, then, might approach such problems in a different manner to school pupils of old - the methods may be less elegant, and provide less scope for creative and penetrating thought - but they offer a more systematic approach.

Consider, for example, the following question (to other aspects
of which we shall wish to refer later):

Two lines are drawn from one vertex of a square to the
midpoints of the two non-adjacent sides. They divide the diagonal into
three segments (see Figure 5.2).

(a) Are those three segments equal?

(b) Suggest several ways in
     which the problem can be
     generalised.

(c) Does your answer to (a) generalise?

(d) Can the argument you used in (a) be
     used in the more general cases?

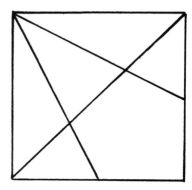

Fig. 5.2

(e) If your answer to (d) is 'No', can you find
     an argument which does generalise?

We shall comment briefly on the purpose of such questions in
Chapter 6. Here we note only that whereas the student of the 1950s had
only purely geometrical ways of tackling the problem, today's student
may well be able to apply algebraic methods to find the solution for
(a). The solution derived by applying a mechanical procedure may be
less aesthetically satisfying than a geometrical one, but are there
other objections to algebraic methods than that of aesthetics?
Arguments against the use of algebraic methods for the solution of
geometric problems have a long history: Simson in the eighteenth
century, we are told, regarded such methods as little better than a
'mechanical knack' in which the student proceeded 'without ideas of any
kind' in order to 'obtain a result without meaning and without being
conscious of any process of reasoning'. Similar objections were used
by those supporting Euclid's geometry in the early nineteenth century,
for example, Ohm in Germany and Whewell in England. They were more
recently repeated by Thom (1973) who pointed out how algebra is rich in
'syntax', but weak in 'meaning', whereas geometry is the reverse.

This difference between the two topics greatly influences
examinations and the attitudes of pupils and teachers. Syntax is easy
to examine and learn, but pure geometry is weak in syntax with the
result that examination questions on it tend to divide between those
which are almost entirely recall and those which are at a much higher
cognitive level. As a result, where examinations loom large, students
and teachers will, given the opportunity, tend to ignore geometry and
concentrate on the 'more profitable' algebra.

Summarising, those planning school curricula for the 1990s would appear to be faced with three alternatives:

Alternative 1.   One abandons the idea that geometry should/can be treated in the school as a <u>system</u> of knowledge (organised deductively or not) where concepts and facts have to be known simply because they belong to the system. Instead, geometry and space are seen as sources of excellent topics for initiating activities of many kinds at a variety of levels. The 'service' aspects of geometry teaching will be met through the provision of algebraic methods.

Consequences:1.   A long tradition will be broken and a part of school mathematics which has fostered a certain type of creativity and imaginative thought through the solution of geometric 'riders' will be lost.

           2.   Since such creativity and imaginative thought are for the very few, the majority of pupils will experience more success if our aims better match their capabilities.

           3.   The teaching of 'proof' at a school level will be severely restricted. (This should mean a recasting of courses at university level – <u>not</u> merely complaints that 'students come to us with no understanding of the nature of proof'.)

Alternative 2.   One still attempts to teach an axiomatic or pseudo-axiomatic school geometry course, based on either 'modified' Euclid (for example, Pogorelov or SMSG) or on, say, transformation geometry.

Consequences:1.   Only the few are likely to be able to cope with the intellectual demands set by such courses, and even fewer will appreciate the significance of the 'game' which they are playing.

           2.   Those few will be well rewarded and introduced to aspects of mathematical thought which cannot be conveyed at a school (or possibly any other) level through any other content area.

Alternative 3.   For some pupils, at least, one presents 'islands' of geometry, i.e. 'local' deductive systems, within the general curriculum. (Thus, for example, one might have units on the angle properties of the circle, and on elementary projective geometry.)

Consequences:1. One attempts to preserve some of the old spirit of
                 school geometry, to utilise the marvellous stock of
                 examples and exercises amassed over the years, and to
                 instil a feeling for proof and deductive chains.

             2. The able young mathematician will still encounter
                 traditional challenges and rewards.

             3. Teachers may ignore teaching such 'islands', since
                 they may not appear to be central to the curriculum
                 and because they will almost certainly cause problems
                 for pupils because of the new modes of thought
                 demanded of them.

3. Applications

In recent years there has been much talk of 'applications', of
making mathematics teaching 'relevant', and of teaching 'mathematical
modelling'. Indeed, in Chapter 3 we have already discussed the place
that applications might hold in the curriculum. Yet there is little
novel in this wish to teach mathematics so that it can be applied: the
first geometry book to be published in English, Recorde's Pathway to
Knowledge (1551), emphasised in its full title that it was intended to
show how geometrical principles might 'moste aptly be applied unto
practise'. Seventy years ago, Carson (1913, p.35) wrote:

'Among the many changes in mathematical education during the
last twenty years ... one element at least appears throughout; a desire
to relate the subject to reality, to exhibit it as a living body of
thought which can and does influence human life at a multitude of
points'.

In fact, historically the teaching of mathematics almost always
has had a strong vocational component: it was only with its insti-
tutionalisation within schools and universities that the links with
reality became neglected or distorted. The reasons for this are
complex, yet still highly significant.

Over one hundred and fifty years ago the 'practical' nature of
algebra teaching was deplored by De Morgan:

"'A person has two horses and a saddle worth £50; now if the
saddle be put on the back of the first horse it will make his value
double that of the second, but if it be put on the back of the second,
it will make his value triple that of the first; what is the value of
each horse?'

Now if all this be meant for improvement in theory, then ....,
but if they be practical questions, we need only say ... that the
gentry have a better way of determining the value of their horses".

This old example well illustrates a problem of teaching applications and of using 'real-life' examples which has still to be resolved. 'Application' within mathematics education has almost as many interpretations as 'knowledge' has levels. 'Applications' can serve a variety of purposes and it is essential when planning a curriculum for the 1990s to distinguish between them.

As De Morgan observed, the question given is not a practical one, yet it might have considerable pedagogical value. The student has to translate a word problem into symbols and then to go through the mathematical act of setting-up and solving a system of linear equations. This is a non-trivial but valuable operation to learn and practise. Is it possible to ensure that all the situations we present to students are really practical and should we attempt to do this? One notes, for example, that much of 'fun' mathematics is based on situations that can hardly be described as 'real-life'. The problem of transporting a wolf, a goat and a cabbage across the river is many centuries old, yet, surely, it was never 'practical'. The key here is the student's attitude to, and acceptance, of the task of solving the problem (see, for example, Christiansen and Walther (1986) for a discussion of task and activity in mathematics education). Whimsical and artificial examples are not necessarily bad, therefore, provided that we do not claim that they are 'realistic'. The art of learning to apply mathematics can be cultivated on other examples than 'real-life' ones.

Another type of traditional example was perhaps too deeply rooted in 'reality'. It is the strictly vocational example, e.g. on elementary book-keeping or stocks and shares, or in some cases linked with the mathematics of a particular local industry. As we shall see below such examples raise questions about 'whose reality' and, in a world of very rapidly changing vocational demands, often prepare students to meet the demands of a past age. The place of such exercises in curricula for the 1990s must be seriously questioned. However, there will still be a need for 'applications' which meet a specific need of society, e.g. taxes, hire purchase and other borrowing arrangements, and (from the Sixth Grade syllabus for Arab Gulf States) distribution of inherited money according to Islamic Law.

Even when one excludes examples of the type we have described, applications can still take a variety of forms. 'Where should a goalkeeper stand to narrow his opponent's angle of shot?', 'How far back from where a try is scored should the goalkicker bring the ball so as to maximise the angle which the goalposts subtend from his kicking position?' Such questions are useful tools in the teacher's armoury. They relate mathematics to actual situations and can have interest and appeal for certain students; geometry becomes less academic. Yet we are not using mathematics to solve a problem; no goalkeeper or goalkicker ever calculates where he should stand or kick from in this manner. Such examples provide links with the world and, in a sense,

legitimise the mathematics being studied, but they are unlikely to
motivate a student to study mathematics. The subject's significance
does not lie in such 'applications'.

        More importantly, the 'link' may not always be with the
'reality' of the student (who may know nothing of rugby football and
its rules). As Carson (op. cit.) wrote: 'The essence of reality is
found in recognisable precepts and concepts, and is therefore a
function of the individual and the time; what is real to me is not
necessarily real to another, and much that was real to me in childhood
is no longer so. It is for the teacher to determine the realities of
his pupils ... . He must also find it necessary to enlarge their
spheres of reality, but he must avoid confusion between a name and a
thing ... [i.e. between presenting them with new names and actually
helping them acquire new concepts]'.

        Carson's essay on 'The Useful and the Real' still repays study
for it clearly shows both the complexity and the long-standing nature
of the challenge of incorporating 'applications' into the curriculum.

        The examples given so far in this section are of 'reality'
being used in the service of mathematics teaching. An altogether
different type of example arises when mathematics is used in the
service of reality i.e. when we use mathematics to solve problems,
rather than reality to provide them. An interesting example of the use
of mathematics to describe, understand and so help solve a social
problem only too familiar to the students is given by Grugnetti (1979)
who considers the hereditary transmission of two of the most serious
illnesses affecting Sardinian people. This example, of course, differs
from those previously given not only in its social import, but also in
that it is not tied to the comprehension or practice of a single
mathematical technique: it draws upon many aspects of the students'
existing knowledge and provides motivation for extending this in
directions which will be immediately applicable.

        Once again it is work carried on in a long educational
tradition: 'Purposeful activity in a social situation' was W.H.
Kilpatrick's description in 1918. The use of projects (Dewey,
Kilpatrick), 'centres of interest' (Decroly) or the 'method of
complexes' (post-revolutionary USSR) to motivate study is an old one.
There have been many recent examples (see, for example, work of
Mellin-Olsen and others described in Howson and Mellin-Olsen (1986)).
Indeed, guidelines set out over 50 years ago for such work are still of
value when contemplating the design of inter-disciplinary projects for
the 1990s:

Does the material come near enough to the student to be real to that
        student?
Does the unit give opportunities for the pursuit of problems, purposes
        and interests on the student's present mental level?

Does the unit stimulate many kinds of activities, creative,
        intellectual and social, providing both for individual
        differences and for the integration of diverse activities?
Does the unit bring about growth from the present level to the next
        step both in individuals and groups?
Does the unit stimulate a desire on the part of the individual to
        proceed on his own initiative and to take responsibility in widen
        his interests and understanding?
Does the unit help meet the demands of society and help clarify social
        meanings?
Does the unit lead to desirable intellectual, social and moral habits
        such as perseverance, co-operativeness, open-mindedness, good
        judgement, self-direction and initiative?
        (see Connell (1980, p.286) for details of sources)

        As we have remarked in earlier chapters, there is now an
increasing wish to see inter-disciplinary, socially-founded project
work introduced into schools. It is important that when advocating
such work we should also look carefully at previous attempts to
introduce it and to try to understand why such efforts proved
short-lived except within a few individual institutions. This type of
work is not easy to introduce into schools: it demands the support of
all teachers and a willingness to readjust goals; it cannot be easily
undertaken within a system which is rigidly governed by examinations
set up on strict disciplinary lines.

        In such projects students see how mathematics can be applied to
genuine 'real-life' problems and, one hopes, gain motivation further to
study the subject. Opportunities will arise naturally for
'mathematical modelling'. Yet here we must distinguish carefully
between engaging in an activity and studying it. M. Jourdain, to his
great delight and surprise, discovered that he had learned to speak
prose, but he had not studied it. Similarly, one can successfully
follow courses in Newtonian mechanics -indeed, many students do -
whilst remaining totally ignorant of the activity of mathematical
modelling in which one is engaged. Actually learning about
mathematical modelling involves a move to a different level, for now it
will be necessary to consider many models and to abstract principles
and procedures embedded in them. (Perhaps Fujita's model is of value
here. For it is part of mathematical literacy to learn about some
standard applications of mathematics; it is part of 'applied
mathematical' thought to develop modelling skills.) It is not always
obvious when calls are made for 'mathematical modelling' in school what
exactly is intended.

        What level of cognitive demand are we making? Exactly what is
the relation between 'understanding' and 'knowing' mathematics and
being confident enough to apply it? How does one help/prepare teachers
to become teachers of modelling? Much developmental work remains to be
done. (What, indeed, are the inferences to be drawn from modelling
curriculum design in probabilistic terms, i.e. awarding some 'weight'

for the successful introduction of a new topic or approach and assigning some subjective probability for the innovation actually reaching classrooms in a worthwhile form?)

A great challenge for the 1990s then is not merely to provide examples of applications which can be usefully studied in schools, but to investigate more deeply the pedagogical purpose and justification of the various types of applications, and to consider why, despite many decades of effort, so little has been achieved within educational systems at large (see, for example, Burkhardt (1983)).

Teaching students how to apply their mathematics must therefore be a major goal for the 1990s. Here three (not mutually exclusive) alternatives exist and demand consideration.

<u>Alternative 1</u>. Mathematics is applied within mathematics lessons.

<u>Consequences</u>:1. Motivation is immediate.

2. Problems will arise in many cases because extra-mathematical knowledge will be required if the contexts are to be meaningful.

<u>Alternative 2</u>. Mathematics is applied within other lessons.

<u>Consequences</u>:1. This can bind together mathematics teaching with that of Physics, Biology, Geography, ...

2. Problems of coordination arise and of matching the needs of other subjects with pupils' mathematical preparedness and maturity.

<u>Alternative 3</u>. Mathematics is applied in inter-disciplinary projects.

<u>Consequences</u>:1. A long-standing educational goal is achieved.

2. Problems arise because of the cooperative working required and clashes with traditional patterns of curriculum organisation.

Yet is learning how to apply mathematics a sufficient goal in itself?

Some would think not. Mathematics now plays an important role within society: how many students <u>appreciate</u> this? Is there, then, a case for helping students to appreciate the uses of mathematics?

Traditionally mathematics education has concentrated upon the learner's acquisition of skills and techniques (products). Now, as we have seen, greater emphasis is being laid on processes. Yet the over-riding goal is still to engage the student in various kinds of mathematical activity. Is there a case, then, to be put for some kind of 'mathematical appreciation'? It must surely be the aim of mathematicians not only to encourage participation in mathematical activities but also to make the populace - and future policy-makers - aware of the vital role of mathematics in national life.

Such work will not fit into normal teaching patterns, neither will it be readily examinable. It would, however, appear to merit serious exploratory developmental work in the coming years.

4. Calculators

Perhaps the major concern relating to the content of the school mathematics curriculum in the 1990s is the extent to which it will/should be affected by new technology and, in particular, by the hand-held electronic calculator and the microcomputer. It is often the practice nowadays to consider these two objects together. There would seem to be advantages, however, in our context to consider them separately. For poorer countries, a reasonable supply of calculators may well be a possibility, i.e. on a scale which enables individual pupil use, whereas micros are unlikely to be readily available even if their price continues to fall. Again, it is the case that the computer opens up new vistas in a way which calculators do not, for it has the potential, through, say, information processing, materially to affect our concepts of both mathematics and how it is applied.

The rapid technological explosion has had many consequences for mathematics educators. Some of these have been remarked upon earlier. Here we note first that it was only late in the day that an open discussion was mounted at ICME 3 (Karlsruhe, 1976) on the effect of the calculator on mathematics education. As a result of the greatly increased accessibility of calculators considerable attention was focussed on them in the late 1970s. Soon, however, attention became concentrated on the use of micros. Considering the vastly wider range of possibilities offered by the micro, and the fact that experimental research and development work is frequently in the hands of comparatively few individuals who work independently, this switch of emphasis is understand-able even if somewhat regrettable. Certainly, the potentiality of the calculator for helping children come to terms with arithmetic has not been greatly exploited; neither has its use for the teaching of more sophisticated mathematical ideas and concepts.

There is no doubt that calculators are now the natural tools with which to carry out arithmetical operations. For that reason learning to use a calculator - and to use one sensibly - must now become part of learning arithmetic. Any attempt to ban their use would be foolish and would only result in alienating students from school mathematics. We need to build up positive attitudes to calculating and the calculator can assist in this. Calculators enable many children to use arithmetic in real situations, including other school subjects, for they facilitate early work with large and small numbers. Access to a calculator enables children to generate number patterns, to explore number properties and to make and test hypotheses, whilst using them makes children focus attention on the order of operations. They aid in the acquisition of the important skills of estimating and approximating. Nevertheless, there is still no consensus on exactly how soon calculators should be introduced and on how their capabilities can be best exploited in early mathematics teaching.

Such research results as are available would appear to be in favour of their early introduction and would not support the theory that use of a calculator is likely to cause a pupil's brain to rot! See, for example, Hembree and Dessart (1986) which is a synthesis of some 80 case studies and indicates that, except for an anomaly at Grade 4, students at all grades profit in achievement and attitude from the use of calculators in mathematics instruction.

It is, however, at the earlier levels of schooling, where one is attempting to lay the foundations of arithmetical understanding, that there is still some controversy and doubt concerning the use of calculators. Here more research and development work is urgently needed.

Further up the school the question is not so much 'How can the calculator assist?' but 'How much time can be spared to explore the many possibilities opened up by its use?'

The teaching of statistics has been completely changed by our ability to deal so quickly with numerical data (and to have in-built programs for, say, calculating standard deviations).

Geometrical progressions, exponentials, infinite series, factorials (and with them combinatorics and probability) can now be more readily studied.

Iteration has become a numerically straightforward operation, for example, studying the limit as $n \to \infty$ of x_n when

$$x_{n+1} = \frac{1}{2}\left(x_n + \frac{3}{x_n}\right),$$

or solving equations by the Newton-Raphson method.

A topic omitted from school syllabuses for many decades because few could cope with the arithmetic involved, continued fractions, now reasserts its claim for inclusion. Children nowadays will find decimal numbers such as 1.047619 more familiar than fractions such as $^{22}/_{21}$.

To go from the latter to the former one merely presses a few buttons. But how can one perform the reverse operation? Such work can help students to conceive of fractions not merely as results of a division, but as special and interesting numbers which sometimes give surprisingly easy and good approximations to real, irrational numbers (π, $\sqrt 2$, $\sqrt 3$,...) .

Procedures for obtaining results provide further subjects for investigation. This can be done at two levels. One is how the operator can best use the calculator, for example, by comparing different key sequences which can be used to generate the values of a particular function such as $x \to x^2 + 3x + 4$. The other is to study the way in which the calculator operates:

how accurate is it?
how does it tabulate functions?

Not all students will be able to attempt these last questions, but their study will bring many mathematical rewards.

Calculators, then, have an essential role to play in the mathematics curriculum of the 1990s. It is essential that more emphasis be laid on studying how best they can be used to enable children better to gain arithmetical understanding.

5. Computers

Never in the history of mathematics education has any development opened up such a vast range of new possibilities and challenges to the educator as has the microcomputer.

It has done this in three ways:

(i)   mathematics itself and the way in which mathematicians work have been affected;

(ii)   as a result of (i) and of the availability of software to carry out a wide variety of mathematical tasks, entire curricula must now be reappraised;

(iii)   new opportunities for teaching and learning mathematics are opened up through the use of the computer as an aid.

The first ICMI Study (ICMI, 1986a) was devoted to the theme of 'The impact of computers and informatics on mathematics and its teaching'. That study was particularly directed at the university and senior high school levels, yet much that was reported applies to other parts of school systems.

Perhaps the actual changes in mathematics itself are less noticeable at a school level, but even there we note that the new methods available whereby a learner can arrive at the concepts of, say, 'function' and 'variable' will almost certainly mean that their meaning will gradually evolve. (We note how, for example, the notion of

'function' has gradually changed over four centuries. That it should
continue to evolve as a result of new uses and applications should not
surprise us.)

       The fact that professional mathematicians now make great use of
computers can, of course, be advanced as an argument for ensuring that
students gain facility with them - particularly since that facility can
be put to very good use at a school level.

       Even, however, if students were not to see a computer until
after they left school, it would still be necessary to rethink
curricula because of the changes of emphases computers have brought
about. Here we mention but three of these:

(a) Algorithms

       Algorithmic processes lie at the heart of mathematics and
always have done. Now, however, there is a new emphasis on algorithmic
methods and on comparing the efficacy of different algorithms for
solving the same problem e.g. sorting names into alphabetical order or
inverting a matrix.

(b) Discrete mathematics

       Computers are essentially discrete machines and the mathematics
that is needed to describe their functions and develop the software
needed to use them is also discrete. As a result, interest in discrete
mathematics - Boolean algebra, difference equations, graph theory, ...
-has increased enormously in recent years: so much so that the
traditional emphasis given to the calculus, both at school and
university, has been called into question.

(c) Symbolic manipulation

       The possibility of using the computer to manipulate symbols
rather than numbers was envisaged in the early days of computing. Now,
however, software is available for micros which will effectively carry
out all the calculus techniques taught at school - differentiation,
integration by parts and by substitution, expansion in power series -
and will deal with much of the polynomial algebra taught there also.
Is it still necessary to teach students to do what can be done on a
computer?

       Finally, we note some of the varied ways in which the computer
can be used as an aid to teaching and learning:

(i)   as an electronic blackboard or teaching aid for the teacher,

(ii)   as a tutorial aid for the student, facilitating the learning of
        new topics aided and assisted by computer software,

(iii)   for individual instruction of students (including 'drill and
          practice' and self-testing),

(iv)   as a tool which will perform calculations, give the values of
          functions, plot functions, ... ,

(v)   as an agent for experimentation and exploration, including, for
          example, learning more about the behaviour of mathematical
          objects by having the computer simulate those objects.

        Given that these are but a selection of the challenges and
possibilities recently opened up by the micro, it is not surprising
that we are still in doubt as we look forward to the 1990s about the
extent to which the computer will/should influence the school
curriculum.

        One basic question which still cannot be answered is the extent
to which we should expect students themselves to develop programming
skills. The more ready availability of software would appear to remove
the necessity for the acquisition of sophisticated skills - at least in
mathematics classes. On the other hand, the greater availability of
micros in the home and the experience which students have gained with
them there has caused people to question with what other types of
learning we can most profitably compare that of learning to use a
computer. Is it comparable to learning:

                    to use a telephone,
                    to drive a car,
                    the mother tongue,
                    a foreign language?

Certainly we traditionally approach the last four types of learning in
very different ways. Yet each would seem to share some similarities
with learning to use a computer. As we move through the 1990s such
problems may be resolved. In the meantime approaches will constantly
have to be modified in the light of the changing experience of
computers which students will have gained outside mathematics lessons.

        What is clearer is that the ultimate goals of school
mathematics are unlikely to be changed by the presence of computers in
the classroom and that computing will not replace mathematics. Yet at
a secondary level there are likely to be changes of curricular
emphases.

        We have already drawn attention to the need for more emphasis
on algorithms (see, for example, Engel, 1977). Although one would not
expect to see complexity theory entering schools, it would seem
appropriate at the senior school level also to draw attention to the
efficiency of algorithms and to distinguish between those, say, with
polynomial and non-polynomial time complexities.

It is likely that there will be pressure for the inclusion of more discrete mathematics at school, yet it is unlikely that this will lead to the abandonment of the teaching of the calculus. Those who wish when they leave school to study the social sciences or biology, say, rather than mathematics, physics or engineering, may, however, be allowed to opt for discrete mathematics modules in preference to ones on the calculus. (See, for example, ICMI (1986a) for more detailed arguments.)

Yet the teaching of the calculus must, surely, change. Already there exists software (see, for example, Tall and West in ICMI (1986a)) which could profitably be employed by any teacher of the calculus. The ability to plot graphs on the micro and to magnify sections until the graph appears linear, or alternatively to see that, however one magnifies it, the graph never appears straight, revolutionises the teaching of differentiability. We cannot afford to neglect such a powerful teaching aid. Indeed, there is no shortage of examples of how the micro has already been shown to assist the teaching of traditional calculus.

To what extent traditional calculus is still appropriate is a much more difficult question to answer. First we note that even if calculus retains its place in the curriculum then, because of the changes brought about by the computer, it will still be necessary to give additional emphasis to its numerical aspects e.g. numerical integration. Indeed, it may well be that discrete mathematics will not find its way into the general curriculum as a distinct entity, separated from the calculus module, but that instead the calculus course will be redesigned and enlarged to include discrete material. An example of a course planned on these lines is described by Seidman and Rice in ICMI (1986a). But if all definite integrals can be evaluated numerically and if software is available for all integration techniques taught at school, what should we now teach about integration?

This question is not easily answered. It may well be that, say, future engineers would find that a somewhat vague notion of integration as being anti-differentiation, some type of sum and/or the area under a curve, might suffice. This hardly seems to be a secure enough foundation though for a would-be mathematician! If the calculator removes the necessity for acquiring and perfecting arithmetic skills, and the micro those associated with polynomial algebra and the elementary calculus, where - and when - is the professional mathematician to gain skills which (s)he will still need?

One attempt to come to terms with this problem is described by Lane, Ollengren and Stoutemyer in ICMI (1986a). There the student learns to sketch the graph of a rational function, by using the micro to find the zeros of polynomials and to differentiate. Of course, it would have been easier just to get the micro to sketch the graph without any recourse to the calculus. However, here we see a deliberate attempt to separate out general pedagogical aims from operations which can be carried out on the micro. We might not agree with the solution with which these authors are experimenting, but their work is indicative of the kind of decisions which educators will have to take in the coming years.

There will be many pressures to cut down on traditional curriculum content in the next decade. It will be necessary that when doing so we pay great heed to the possible consequences of our actions. Certainly, there is no doubt that the use of the computer can improve students' understanding of the calculus and of other parts of mathematics (for example, statistics and probability where the gains from being able to simulate are enormous, or the study of geometric transformations). It is not yet clear though that this will lead to any appreciable saving in the time taken to teach topics – indeed, there are strong hints that in general there will be no time saved, simply the promise of better understanding.

The use of the computer as a teaching medium – i.e. for the purposes of individualised instruction – will be considered in the next chapter. More immediately relevant to the content of the curriculum is the use of the micro to encourage experimentation and investigation.

As is pointed out in ICMI (1986a, pp 22–23) exploration and discovery are important components of the educational process of mathematics and are greatly facilitated through the use of computer technology. Two, three, even four dimensional objects can be investigated with the help of computer graphics, as can such interesting geometries as 'flatland' and turtle geometry. Different algorithms can be designed and executed for similar or related tasks. The inductive paradigm – compute, conjecture, prove – can be employed in many different situations. The list could, of course, be made much longer.

Moreover, much work to which we have already referred, for example, the use of spreadsheets and CAD packages, has still to be properly developed and tested, yet would seem to possess tremendous potential for influencing mathematics teaching as the 1990s progress.

The authors of the ICMI (1986a) report allude to three key problems associated with this type of working. One, the preparation of teachers to work in this new mode, is a perennial problem which applies to any significant innovation in teaching procedures. The other two relate to the selection of suitable tasks or situations for the students to explore, and the difficulty not only in testing, but also in recognising, reinforcing and consolidating what the students have learned.

Certainly, it is not always easy to find suitable tasks for students. Simply having them sit in front of a micro is not sufficient. Finding the largest perfect number one can with the micro's aid is <u>not</u> a sensible mathematical task. Programming the computer to detect a perfect number is, but sitting there whilst the micro churns on is not. Again, 'investigating' for different values of x_1 the behaviour of the sequence defined by $x_{n+1} = x_n/2$ if x_n is even, $x_{n+1} = 3x_n + 1$ if x_n is odd, is unlikely to prove of much value. Yet we have seen both of these tasks suggested. Perhaps the prime aim of investigations is to train students to ask sensible questions. In the first case, they are presented with a weak mathematical question as a model, namely 'Which is the largest perfect number you can find?', in the second, whatever sensible question they pose, they are unlikely to have any success in their attempts to answer it.

Again there is a risk that the computer will be used as a steam-roller which, by producing its results so quickly and ham-fistedly, makes it more difficult to appreciate the true nature of the problem. For example, an interesting non-computer-based investigation for children is determining which totals cannot be made using, say, only 5p and 7p stamps (see, for example, Ahmed in Damerow <u>et al</u>. 1986). A micro can, of course, provide a print-out of obtainable numbers in seconds. As Atiyah writes in ICMI (1986a) of a similar kind of situation occurring at a higher educational level: 'Is this to be the way of the future? Will more and more problems be solved by brute force? If this is indeed what is in store for us should we be concerned at the decline of human intellectual activity this represents, or is that simply an archaic view-point which must give way before the forces of "progress"?' (p.48).

These examples all give rise to the key question 'What happens when the machine is switched off?' (see Mason, 1985). Given a list, say, of the first four perfect numbers does one look for possible patterns and even possibly conjecture what a fifth number might be? Does one try to account for the results with 5p and 7p stamps and to make a hypothesis about, say, 7p and 9p stamps? Or are the results displayed on the screen, recorded and the matter closed?

If the answer to the last question is 'Yes', then what has been learned? Now it can very often happen, say in an experiment with LOGO, that much has been learned: for example, the child may have developed a particular strategy for doing something, and that the actual learning – the hypothesising, adjusting, reworking – is not immediately obvious in the final result. How can the teacher help fix and reinforce what has been learned? This is by no means an easy question to answer, yet unless we attempt to come to terms with it then most of the seeds sown by computer-based investigations will fail to germinate.

The use of computers for exploratory, experimental work is likely to prove difficult and it will disturb the normal teacher-student relationship. However, the potential rewards are great indeed, and further development work is needed, <u>as is detailed evaluation and assessment of existing work</u>. Good software does exist. In some countries, however, it is hard to obtain since the commercial

production of software intended for schools is not particularly
profitable and so distribution of both information and software tends
to be haphazard.

The whole problem area of how computers can and should
influence mnathematics in schools is, therefore, very open. We must
attempt to steer a course between simply using the computer to help us
meet traditional goals, and, in the other direction, abandoning
valuable old practices in search of new, perhaps illusory, goals. A
blend of daring and circumspection will be required. It is essential,
then, that in the coming years there is extensive international
cooperation and exchange of ideas and information in this particular
field.

Chapter 6

Classrooms and Teachers in the 1990s

1. The Classroom in the 1990s

        What will a classroom look like in the 1990s - how should we
like it to look?

        The first part of this question can be readily answered - a
wide variety of ways of organising classrooms will doubtless be found
in the 1990s. But what will be the trends; what would teachers,
educators and others wish to see; and what is likely to prevent
desirable changes from happening?

        In the last chapter we remarked upon the divergences between
the curriculum as intended and implemented. The disparities which
exist in relation to content are, however, small compared to those
concerning teaching methods. For, despite what is urged by educators
and in official reports, and what can be seen in the classrooms of some
outstanding teachers, experience suggests that in the bulk of secondary
school classrooms one is likely to find a stereotyped form of teaching
which relies heavily on the textbook and the traditional teaching
pattern of exposition-examples-exercise. Apparatus is rarely used, and
class-teaching is still the norm. Traditional teaching exerts more of
a stranglehold the further up the school one goes and the more
specialised the students. At least, it prepares students for
university!

        Yet, of course, there are many classrooms in which traditional
teaching is not the norm, and in which students spend much of their
time working individually or in groups. It must be emphasised here,
though, that completely individualised work can be just as formal as
'chalk and talk' lessons. One recalls dreadful programmed-learning
laboratories/classrooms of the 1960s with students sitting at
individual machines like people in an amusement arcade from which every
hint of amusement or interest had been removed. Is our vision of the
1990s to be of students each sitting at an individual micro - all
neatly arranged in rows?

2. Resources and Pressures

Schools are likely to be under considerable pressure in the
coming years to demonstrate that they are technologically aware. The
impact of technology on society is likely to be so great that many
outside education will see it as the solution to most educational
problems. We have already ample evidence of governments pressing
micros upon schools, without any serious thought about the particular
problems which their presence might resolve: their actual presence in
the classroom will be sufficient it is thought/hoped to work some kind
of miracle. Enormous claims are made on the behalf of technology in
the classrooms - claims which are strengthened by examples of what can
be done by enlightened educators taking full advantage of, for example,
the marvellous facilities offered by today's computers. Yet there is a
danger that the good will be swamped by the indifferent and the
downright bad. Much of the software currently on sale can bring little
cheer to the educator. Significantly much educational software was
originally developed not with the school in mind, but the home. In
the past, educational software - in the form of textbooks - has been
written with the teacher or educational administrator as the target
purchaser. Suddenly the bigger and better funded market was thought to
be the home, and educational software produced for it was often tied to
the achievement of easily describable, cognitively low-level aims
(usually linked to examinations and the 'basics'). All too quickly,
however, the attractions of the home market, apart from popular games,
seem to be fading.

In general, switches in target purchaser affect both the kind
of mathematics presented and the way in which it is done. New
expectations concerning content and style of presentation are aroused
in students: expectations which schools might oppose or be unable to
meet.

Such commercially-produced software can also provide
unfortunate and understandable reactions amongst researchers and
developers in mathematics education. As we observed in the last
chapter, and will comment on further in the next, research and
development work in the West can often be unplanned and uncoordinated
and rely upon individual enthusiasms. Thus, as we have noted,
developmental work on the calculator tended to be pushed to one side
and forgotten once the micro came along. Now, certain classroom uses of
the micro, for 'drill and practice' and 'self-testing', are in danger
of being associated with some of the poorer commercial material, and
developers are drawn to the unspoiled, more exciting world of
experimentation and simulation. As was made clear in the last chapter,
this is an educational use of the micro which merits great attention,
but it should not take up all the energy of the dedicated and gifted
developers. The provision of less exciting, more mundane types of
software would benefit enormously from their expertise and commitment -
and the support it would give to the average classroom teacher would in
the short run probably be far greater. Certainly, the 1990s will not

see the computer usurping the place of the teacher. It is now clear
that the claims made in the 1960s for computer-based education were
vastly overblown. Yet if the computer cannot replace the teacher
altogether, it can provide a valuable resource for independent learning
and reinforcement on a limited scale. There are signs, particularly in
the primary age-range, that this is being realised.

To sum up, then, it is not envisaged that in the mathematics
classroom of a developed country in the 1990s we shall see each student
working away at an individual terminal. A more profitable aim would be
to see two or more micros in each classroom as one of a range of
resources for students' learning, together with one for the teacher
linked to a blackboard-size, high-resolution screen. Of course,
provision of computers on this scale envisages a change in the way
classes are organised. It would be difficult to fit two micros into
the traditional pattern of class teaching!

Yet, in addition to the micros in the mathematics classroom, it
is also very likely that schools of the 1990s will possess a computer
laboratory. What part will mathematics have in that? Originally, such
laboratories were often managed by mathematics teachers. Now, however,
the fact that computers have implications for education far beyond the
bounds of mathematics has been generally accepted and specialist
teachers of computer science and information technology are entering
schools. The subjects do indeed have a separate existence from
mathematics, yet it would be most unfortunate if attempts were not made
to coordinate and relate work which takes place in the computer
laboratory to that in the mathematics classroom. Here is yet another
new challenge to be faced in the 1990s.

Other classroom resources may well include video tapes and
discs, which have yet to be successfully incorporated into schools
although marvellous examples are to be found, for example, in the
output of the Open University in Britain, and, of course, written
materials including the familiar textbook.

The coming of new printing and copying facilities within
schools has materially affected the position of what has traditionally
been the major medium through which new ideas and teaching methods have
been disseminated - the textbook. Limited educational budgets have
restricted the buying power of schools - this at a time when
publishers' production costs have, in many countries, soared. The
result has been to make publishers extremely cautious about producing
innovative, and therefore commercially risky, texts. Money spent on
trials cannot be easily recouped. As a result the textbook (in which
one includes published classroom materials as a whole) has developed
surprisingly little in the last two decades; visually there have been
advances, but at a deeper level there have been few changes of
significance. What is the future for the textbook in the 1990s? New
technological developments mean, for example, that it would now be
technically feasible for schools to construct and print their own

textbooks, making use of a data bank of licensed material and programmed 'trees' of prerequisites. Will the textbook and other textual materials have changing roles to play in our teaching? How can we ensure that the effectiveness of textbooks can be increased? What research is required in this particular sphere? Is there nowadays an irreconcilable conflict between commercial publishing and educational needs? Can textbooks be written so as to be informative and supportive, but not prescriptive?

We envisage, then, that the teachers will face opposing types of pressures. On the one hand, it is likely that there will be pressures from various administrative directives which may lead them to increase their use of exposition, recitation and drill as they try to cope with new demands. On the other hand there will be expectations from students, parents and others that they will make use of the new technology – which cannot be successfully incorporated into such traditional teaching patterns.

The latter form of pressure may well tempt educators to be pre-occupied with technology and attempts to fit it into existing educational patterns. As a result the production of radically new material and the serious consideration of new educational ideas and alternatives could well be inhibited. How can we guard against commercial and external pressures over-riding educational criteria?

The problem in developing countries will be of a different nature. In these the provision on a wide scale of expensive technological aids such as micros will not be possible. Developing countries will, then, be prevented from following the lead of the developed ones. Immediate consequences will be the need to define a path which the developing countries can follow and to counter the loss of morale which will occur if it is believed that such technological appliances are now essential for successful mathematical education. Should countries in which the general populace is technologically impoverished even consider introducing technology into the mathematics classroom? Will it be this factor which essentially shatters the hold of a canonical mathematics curriculum?

3. The Teacher's Role

Bearing in mind the points raised in the preceding sections, there would appear to be three contrasting views relating to the position which the teacher might occupy in the classroom of the 1990s.

Alternative 1. The teacher retains the traditional role of 'supreme answer giver', individually (and solely) supplying motivation, explanations and assistance for the class, and rapidly adjusting to its moods and the varying demands of the subject.

Alternative 2. The teacher becomes a guide to learning, and the
               designer (possibly together with colleagues within the
               school) of a curriculum which makes use of a variety of
               different resources – micros, booklets and other written
               materials, peer group interest and assistance.

Alternative 3. The role of the teacher changes to that of administrator
               of a multi-resource learning kit which, it is hoped,
               will carry the main instructional burden.

The attractions of Alternative 1 are that most teachers and
educators can recall being profoundly influenced and inspired by
someone who worked in that mode. What is often difficult to recall are
the changes which have taken place in social and educational contexts
and in the demands now placed upon the teacher, who in the coming
decade is likely to be less well qualified and whose status both
outside and inside the classroom has changed considerably.

The second alternative demands new skills from the teacher, and
an acceptance of a new role in the classroom. A move towards a more
open pedagogical style in which the teacher becomes a helper in the
task of building up the student's own mathematical knowledge requires a
major shift in the teacher's view of what is appropriate and possible.
Such a shift entails helping students reflect on their mathematical
knowledge, becoming aware of what they know and how they learn.
Activities in which students engage in investigations, discuss their
ideas, and write up what they have found may help them become more
reflective. Teachers who have not conducted such activities will need
to see examples and to have their own opportunities to experiment and
reflect. They will also need help and encouragement when making their
first hesitant and insecure steps. For insecurity would seem an
intrinsic part of such teaching. Problem-solving, for example, is a
lengthy, demanding and risk-taking business for both students and
teacher in a way that working through a set of exercises is not. The
teacher must gain the confidence to live with a degree of insecurity.

The third alternative is generally advanced as a response to
the problem mentioned above, the scarcity of well-qualified teachers.
The consequences of such a policy which would almost inevitably
partition schools into two sets – those able to offer teaching based on
alternatives 1 or 2, and those who cannot – have yet to be carefully
assessed.

4.   Allowing Teachers to do a Better Job

Just as one should not view the student as an empty vessel to
be filled with mathematical knowledge, so it would be wrong to see the
teacher as an empty vessel to be filled with pedagogical knowledge.
Most experienced teachers already have considerable pedagogical
knowledge, and many would wish to shift to a more open teaching style:

yet they still might not change. As one teacher remarked, "I know how to be a better teacher of mathematics than I am". Careful consideration needs to be given to the conditions of teaching that prevent teachers from implementing reforms.

A teacher who has to teach 5 or 6 classes a day has little time to reflect on his or her teaching, to prepare instructional materials, to inspect a variety of alternative approaches, or to work with colleagues in developing and renewing a coherent programme. A teacher who lacks a work space, equipment, a budget for purchasing materials or travelling, discretionary time for working with colleagues or students, cannot be expected to develop far as a 'reflective practitioner'. When teachers are further harassed by lunchtime duties, administrative chores and administering mandated tests, then morale deteriorates. If their status in society is also threatened, then so much the worse. To some extent this is a problem shared with teachers of other subjects, but that does not remove the issue as a source of concern. Indeed, the opportunities that dissatisfied mathematics teachers have to enter better-paid, higher-status fields which require mathematical expertise, make the retention of experienced and capable mathematics teachers an especially severe problem.

It would be reassuring to think that the 1990s will bring a lifting of pressures on teachers. However, it must be realised that many of the pressures on teachers are not removable; teachers in many countries must just accept them and do what they can with the little amount of freedom that remains.

Meanwhile it is vital that administrators and researchers should pay more attention to the identification and elaboration of those pressures that cause teachers to 'close up', be unadventurous, and resort to giving more exposition and more drill, etc. To what extent are teachers actually affected by, say, the technical difficulties of managing heterogeneous classes, a lack of mathematical qualifications, the emphasis by administrators, parents and others on measurable achievements by students, the fear of being held accountable for student failure and so on? In-service teacher programmes might well invite teachers to say what obstacles they experience to teaching in the way(s) they would prefer. Sometimes this can draw attention to the fact that not all obstacles are as insurmountable as they may at first appear.

5. Examinations and the Teacher

Pressure resulting from the assessment of pupils' performance is not new, but it is now taking new forms. External examinations were introduced largely in an attempt to control teachers and teaching and this is still one of their major purposes: their existence exemplifies a lack of trust. To the weak teacher they were a crutch, since they helped set what were reasonable standards and provided the teacher with a form of 'self-assessment'. As our conception of what is desirable

knowledge has changed, then increasingly it has become clear that
conventional external examinations no longer test all that we should
like to test. For example, they rarely include 'open' questions such
as that on geometry given in the last chapter (for finding a 'reliable'
marking scheme for such questions is extremely difficult). In some
countries there has been no obvious response to this problem. In
others the responses have differed considerably, but in a way that very
much influences classroom behaviour throughout the year.

        The response in Australia has been to make the assessment of 16
year-olds entirely school-based. Teachers are provided with detailed
guidelines on the evaluation of school programmes and the assessment of
students' performance and within individual schools they then take
responsibility for the awarding of Secondary School Exit Certificates.
In Britain extra responsibilities have been given to teachers in an
attempt to widen the scope of 16+ examinations, but control is still
exercised by the external examination boards. An enormous bureaucratic
system attempts, at no little expense, to guarantee some measure of
comparability.

        In both cases the teacher has to shoulder new responsibilities,
although in Australia some traditional ones have been removed.

        Just how much responsibility a teacher should take for pupil
assessment and certification is not immediately obvious. Of course,
national traditions and expectations vary. For example, Germany has a
long tradition of school-based assessment. Yet this would seem an
important area for research: not merely, as is currently being done, to
develop new techniques for assessment, but as hinted at earlier, to
discover what pressures teachers feel exposed to under the various
systems, and to investigate how teachers respond to the challenges of
being given a greater part to play in student assessment, and to being
recipients of increased trust.

        Summing up then, three key positions can be held on
examinations:

Alternative 1.   External examinations are essential to uphold
                 'standards'/ set goals for teachers and
                 students/satisfy external demands/preserve uniformity
                 throughout educational systems. Student assessment is
                 best carried out externally.

Consequences:1. Examinations will carry out their intended role and
                 will effectively control the curriculum.
            2.   Desired changes in the curriculum will have to be
                 accompanied by changes in examination procedures, and
                 curriculum development will, of necessity, be
                 circumscribed by possible developments in methods of
                 student assessment.

3. Teachers will view examinations as a 'mechanical contrivance' designed and imposed by others which it is their job to help students to 'overcome'.

Alternative 2. Student assessment should be left to teachers.

Consequences:1. There will be a need for well-detailed criteria to be suppplied to schools in order to ensure comparability of aims and performance.
2. The teacher will be awarded greater professional standing, but will have to carry additional work burdens.
3. In the initial changeover period from externally- to internally-moderated examinations there will be new pressures on teachers because of changes in their traditional relationships with students and parents.

Alternative 3. Responsibilities for student assessment should be shared between teachers and external agencies.

Consequence: 1. Care must be taken to ensure that internally assessed work is not given less weight than the externally assessed component. Otherwise, the burdens on teachers will have been increased with little positive gain.

6.   The Teacher as 'Victim'

It is very easy to go on adding extra loads on teachers, by giving them, say, greater responsibilities for curriculum design, choice of teaching methods, and student assessment. Clearly, these are all educationally desirable if the teachers' backs are strong enough to carry such weights. We remarked in Chapter 4 that, so far as content was concerned, in many countries 'less' was being demanded of students in the hope of achieving 'more'. Yet everywhere there seem to be increasing loads on teachers. Do we, then, run the risk of 'blaming the victim'? The Governor of a Southern state of the USA is reported as having replied to criticisms of his state's notorious prison system by saying 'There's nothing wrong with the prisons. We just need a better class of prisoners'. We must beware of falling into the same trap when talking about teachers (and students). Proposals should not be formulated in terms of teachers we should wish to see in schools, but within the capabilities of those that are there. The conditions that prevent teachers from implementing desirable changes in practice must not be ignored. Moreover, when looking ahead to the 1990s, it must never be assumed that teachers themselves have no role to play in analysing their situation and finding a means of change.

Chapter 7

Research

1. The Status of Research

There has been a considerable increase in the amount of research carried out in mathematics education in recent years. This is evidenced in the number of research degrees awarded, of centres established, of journals published and of national and international meetings held.

In many countries, however, the peak for spending on educational research has now passed; in the developing countries little money has ever been expended on research activities, and even in those countries thought of as 'big spenders' the actual percentage of the educational budget allotted to research and development has always been miniscule. Thus, for example, in the USA investment on research and development accounted in 1983 for less than one-tenth of one per cent of the total expenditure on education: comparable figures for defence, health and industry were 15, 2 and 2.6 per cent respectively (N.R.C., 1985, p.43).

Why is educational research and development so poorly-rated? Why, when educational systems devour so great a proportion of national budgets, is so little money spent in studying their working and in developing ways to improve them? The proportion of children whose lives are blighted by their 'education' must be considerably higher than the proportion of adults who die from some of the medical conditions which will always attract research money.

The cynic will answer that educational research suffers because of educational researchers; that their low esteem is a natural, and just, recognition of their past lack of effectiveness, and that the whole educational process is too complex to permit sensible analysis. That is, the researcher, tackling an apparently intractable problem, is also presented with enormous problems of identity and status.

Is it the case, then, that the researcher has nothing to contribute to the solution of the problems which school mathematics will face in the 1990s? Surely not! Will it be necessary, though,

for research activities to be given new focuses and for the present
organisational framework for research in mathematics education to be
improved?

2. Forms of, and Frameworks for, Research

     Let us begin by considering the latter part of the question
with which we ended the last section. To an enormous extent, the style
of research in education is governed by the context in which it is
undertaken and the manner in which funding can be obtained. Notable
consequences in the developed Western countries are the pressures to
'publish' and the part played by 'Ed. D.' and 'Ph.D.' research,
undertaken with the major aim of acquiring a qualification and, en
passant, a training in how research might be undertaken. The
limitations of the findings from the latter type of research are all
too evident. When 'biological type' investigations are carried out,
then populations are frequently too small and socially and geographi-
cally biased for reliable conclusions to be drawn; on the other hand
'anthropological-mode' research makes demands on the experience of the
researcher which are rarely met by Ph.D. students (that they can be met
is demonstrated in Erlwanger's (1974) influential thesis). In general,
it must be assumed that valuable educational research is much more
likely to be done by trained and experienced researchers, even more so,
because of the multi-faceted nature of education, than is the case in
other fields of scientific and literary enquiry. In this context,
centres for research in mathematics education – the IREMs in France,
the IDM in West Germany, the Shell Centres in England, those at the
Universities of Georgia and Wisconsin in the USA,... – will clearly
have a central role to play. Yet what is that role to be?

     The first option, and one that might prove particularly
suitable for a research centre in a small and/or developing country, is
that it should be the national agent for organising and carrying out
research. The country then has a clear focus for research activity,
professionalism can be acquired and, with it, greater esteem for
researchers and their output.

     The second option is that the 'centre' should not only carry
out its own research, but should actively seek to promote and stimulate
others, not career-researchers, to engage in research also. The risks
of misinterpretation here are obvious: if anyone can engage in
research, then what status can researchers hold? What does it mean to
talk of a teacher-researcher?

     Here it is important to look at the levels at which educational
research can be carried out and the differing audiences to which it can
be addressed.

On one level we have research work intended to help decision-making by administrators and curriculum designers. Thus, much of the work of the Second International Mathematics Study (see, for example, Crosswhite et al, 1985) or of the Assessment of Performance Unit (APU, 1984) in England falls into this category. (One notes here the ancillary project (Quadling,1986) established to search in the APU's findings for implications for teachers.) Such work always, of necessity, requires large resources and skilled personnel. It is usually funded by Government agencies and, for that reason, has in recent years been increasingly tied to the assessment of pupils and to other evaluative procedures to check whether or not public money is being expended wisely.

A second level of research can be seen as research carried out for researchers and theorists. A weakness of much current research is that investigations are too often divorced from any theoretical context; they do not aim specifically to develop theoretical constructs and, as a result, do not help to fill that great need of mathematics education, the development of theories for practice. Kilpatrick (1981) draws attention to the need for conceptual, theory-building analyses and also offers a useful model of research in which he looks at the two dimensions, rigour (the professional standards which research should meet) and significance (the extent to which results might influence thinking). He contrasts the high-rigour low-significance of much US research with the low-rigour high-significance of many Soviet studies, and asks how we can move to a state of high-rigour high-significance. Here, again, it is necessary to stress that the multi-faceted nature of mathematics education necessitates strong inter-disciplinary research teams. Once again we see the benefits of centres having strong links with researchers in other disciplines or being sufficiently large to include on their staff workers drawn from allied fields.

The immediate influence of such research on teaching is, however, likely to be limited. Research will have an impact on the practice of teaching if and only if it produces, or leads to the production of, appropriate 'techniques' which can be adopted in the classroom. 'Fundamental research' (whatever that might be in education), new theoretical constructs, phenomenological studies, etc. will not directly affect what happens in classrooms and it would be wrong to expect them to do so. The research 'results' so produced are of interest mainly to other researchers and to curriculum developers. It is the latter who must ensure that these results become incorporated into 'tangibles' – methods, materials, apparatus, etc. – which can be adopted by practitioners. (Here, we do not rule out that gifted individual teachers will make the necessary translations.) In this connection, the practical importance of theories must again be stressed: properly handled a theoretical rationale allows the reader, be he developer or interested teacher, to see beyond the specifics of the research study under consideration to the general constructs that lie behind it. This in turn will enhance the study's generality and its applicability to practice.

At yet another level, 'research' can be the concern of the
teacher, of the 'reflective practitioner'. To paraphrase Freudenthal,
the teacher should explore what is known about learning and about what
can happen in the classroom and try to use this knowledge in his/her
teaching. The results of such 'research' may vary in quality and in
general applicability, but the process of reflecting is likely to
enrich both the practitioner's teaching and his/her self-esteem as a
professional. It is not the type of research which is likely to appeal
to funding agencies - but, in any case, it needs time rather than money
for pursuit. Yet teachers must realise that we all have within our
experience a vast amount of data about learning, doing and teaching
mathematics. Reflection is the technique for unlocking what is stored
and scrutinising it. The procedure is 'subjective' (but all research
involves subjective judgements somewhere) and requires validation by
observing or provoking students' mathematical behaviour.

Of course, these three 'levels' are no more than rough
labelling devices and certainly are not meant to imply that the
'career-researcher' should be separated from the 'reflective
practitioner'. Indeed, a key question for the 1990s is:

How can teams of teachers and researchers be organised to
address research problems of mutual interest and concern, and what
techniques are effective in sustaining the operation of such teams?

3. Foci for Research

Yet what should be the focus of research in the immediate
future? What are the key questions to which researchers should address
themselves - bearing in mind that the problems should be reasonably
tractable and that the answers should materially affect mathematics
education in the 1990s? (See Wheeler (1984) for a description of a
small, informal enquiry into research problems judged to be significant
for the future development of mathematical education.)

It is not immediately obvious who should formulate such
questions: administrators, researchers or teachers will be looking for
different kinds of answers. The danger is that the questions will be
formulated de facto largely by the funding agencies. Yet it is
important that mathematics education should share the freedom of other
research areas, namely to establish and to follow 'internal' laws of
development, through the building up of workable research methodologies
and paradigms.

At present, much research is directed at student learning and
behaviour -indeed, it could be claimed that most current research looks
at the learning of mathematical concepts and principles. Recently
there has been more interest shown in classroom studies, of teaching
styles and of teachers' cognitive attitudes, e.g. the way in which
teachers' instructional decisions depend upon the theoretical

frameworks they have developed regarding mathematics and its teaching (see, for example, Christiansen et al, 1986). Much more remains to be done, however, in these and related fields (see, for example, Romberg and Carpenter, 1985).

There is still relatively little emphasis on students other than on how they learn. Should more attention be paid to their attitudes to mathematics and to schooling in general (and to how these are formed), their expectations, perceptions and misapprehensions, and their motivating drives?

Our knowledge of the processes of change is scant: yet improvements will hinge upon the successful implementation of new strategies, new conceptions, new visions. For example, what techniques are effective in persuading teachers to change their beliefs, and to adopt and use new teaching procedures, such as practical or investigational work with which the teachers themselves may be unfamiliar?

If research is to be effective then, as Figure 7.1 shows, it will have to be carried out with a wide range of objectives in mind. This figure (which is an elaborated version of one to be found in N.R.C. (1985)) can still only provide a rough indication of the scope of possible, and desirable, research activity. For example, even in the relatively well-researched area of students' learning one can find large and under-researched fields such as:

(a)   research on spatial abilities: these are so often assumed fixed and little interest has been shown in how they might be developed in their relation to mathematics.

(b)   research on memory abilities: these too have often been assumed fixed and little interest has been shown in activities which might improve memory and in the role of memorisation in the learning of mathematics.

(c)   research on 'meta-cognition': how do students control and direct their thinking, keep track of what they have done, understand their own thinking processes? Meta-cognitive skills appear to be important when students apply their mathematical knowledge to new situations, yet they are given little attention by teachers. Can such skills be developed by teaching and, if so, how?

(d)   research on testing and evaluation: the effect on students of current practices – particularly the effects on learning of short answer testing and of frequent assessment, and the knowledge students have that is not measured or is distorted by tests.

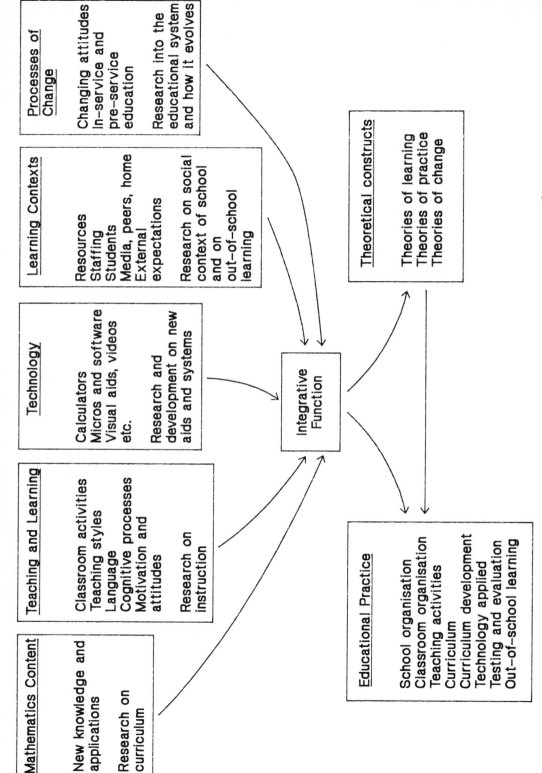

Processes of Change

- Changing attitudes
- In-service and pre-service education
- Research into the educational system and how it evolves

Learning Contexts

- Resources
- Staffing
- Students
- Media, peers, home
- External expectations
- Research on social context of school and on out-of-school learning

Technology

- Calculators
- Micros and software
- Visual aids, videos etc.
- Research and development on new aids and systems

Teaching and Learning

- Classroom activities
- Teaching styles
- Language
- Cognitive processes
- Motivation and attitudes
- Research on instruction

Mathematics Content

- New knowledge and applications
- Research on curriculum

Integrative Function

Theoretical constructs

- Theories of learning
- Theories of practice
- Theories of change

Educational Practice

- School organisation
- Classroom organisation
- Teaching activities
- Curriculum
- Curriculum development
- Technology applied
- Testing and evaluation
- Out-of-school learning

Fig. 7.1

In general, surveying the present scene we see the extent to which research interests and methodology are dominated by the way in which research funding operates - for what are cheap to investigate are students at work; what are easy to provide are data. Studying teachers, for example, is both more expensive - there are not so many around - and less easy, for their consent is required; interpreting data demands experience and considerable understanding.

What changes are desirable in the way in which research in mathematics education is carried out and funded?

Is greater centralisation of resources and/or project planning required? Can international cooperation be more advantageously employed?

Yet another matter of concern is the way that, at the present time, research is largely a badge of affluence worn by the wealthier countries. The needs of the developing countries are, however, just as great. Too frequently such countries import research findings established in educational and social contexts totally different from their own and hence of doubtful validity in their new surroundings. How can developing countries be helped?

It would, however, be wrong to paint too gloomy a picture of the state of research in mathematics education. Great progress is being made as is evidenced by the work of such groups as that on the Psychology of Mathematics Education (PME). Research paradigms are becoming more stable and more generally accepted; data pools are being established and exchanged. There are exciting new developments in inter-disciplinary work in, for example, artificial intelligence and cognition theory. The results are unlikely to have much affect on mathematics education in the 1990s, but could well help shape practice in the following decade.

Continued and increased educational research, then, is essential, as is planned and well-monitored developmental work. At the present time, education finds itself in a similar position to medicine in the nineteenth century. Its practitioners are seen, not so much as scientists pursuing a well-defined discipline, but as possessors, in varying degrees, of 'know-how'. The hoped-for transition of status, to that analogous to late-twentieth century medical scientists, will not be achieved without a considerable strengthening of educational research and theory.

Chapter 8

The Processes of Change

The preceding chapters have been largely devoted to 'change':
to the need to react to societal and technological changes, to foresee
and prepare for change, to make changes. We have been largely
preoccupied with the general question, 'What changes in school
mathematics are required?' Answering such a question becomes, however,
merely an interesting intellectual exercise, unless we also ask - and
seek to answer:

How are changes to be effected?

1. How schools change

There is no doubt at all that schools change. They are not the
same institutions that they were twenty or forty years ago: pupils'
behaviour, awareness, perceptions and attitudes have changed, and so
have those of their teachers. This must be true to a greater or lesser
extent in all countries. Sometimes, but not always, the organisation
of schooling and schools has changed and together with it, curricula.
Change, then, is constantly taking place even though we cannot always
account for it. Yet often, amidst all the changes, the very things we
should most like to change - be they attitudes, practices, content -
remain obdurately fixed. Moreover, schools have changed more slowly
than the societies in which they are embedded.

Our knowledge of how to achieve desired changes, of the way in
which the processes of change are generated, operate and can be
directed, is extremely limited. As we remarked in the previous
chapter, very little attention has been paid to the key topic of how
educational systems change: of how new ideas are taken up, or rejected
by, schools; of the speed at which systems can adapt to change; of the
relative merits of the various strategies that have been adopted for
securing change. Some important work was carried out by the
Organisation for Economic Cooperation and Development in the early
1970s, and in addition to several volumes of Case Studies of
Educational Innovation this also led to the production by
Per Dalin (the Director of the OECD Project) of the important and
significantly named book, Limits to Educational Change (1978). An
attempt to set Dalin's findings, and those of other sociologists, in

the context of mathematics education was made by Howson, Keitel and Kilpatrick (1981). Other important studies of that period were ICMI (1978), which considered the way in which the aims of the 1960s had, or had not, been realised in a number of different countries, and Steiner (1980), which considered the general problem of 'stability and change' in the mathematics curriculum. Since that time very little that is new has been added to our understanding of the processes of change and, perhaps not surprisingly, there has been little further research carried out in the area.

Yet it is not new for change to be sought in mathematics education: the subject has a long history - Lacroix, Pestalozzi, De Morgan are well separated from us in time, yet even they acted not as originators but within a tradition. Changes, too, have been successfully accomplished. Huge educational systems have been established where little existed before, for example in the USSR. Yet planned change is becoming increasingly difficult, for as a result of past efforts, educational systems have expanded enormously in size. Planned educational change now depends upon thousands of individuals acquiring new competencies, accepting new values and goals, and frequently also accepting alterations to the balance of power. It cannot be undertaken cheaply. De Morgan, writing in the mid-nineteenth century before England had a state system of education, was trying to persuade individuals to change. Nowadays there are literally millions of primary school teachers in India - 'curriculum development' has acquired an entirely new dimension. How are teachers in India, or elsewhere, to be helped first to perceive a need for changes, and then helped to make them?

2. Some difficulties of effecting change

The difficulties of bringing about changes in an intended, ungarbled form have been made manifest on numerous occasions in the recent past. As a result, it is now relatively easy for those involved in curriculum development to produce lists of identifiable barriers to change. What is not so easy is to formulate strategies for surmounting those barriers. Yet whilst educators might recognise the difficulties, it is not always clear that politicians do.

Well-intentioned attempts to reform educational practices, for example, the introduction of new examining practices at 16+ in England or the redesigning of the curricula for the final years in French secondary schools, may well founder - or, at the very least, result in minimal improvement - simply because the time-scales have apparently been determined more on the basis of political expediency than on past educational experience of effecting changes. Curricula cannot be amended as readily as laws, nor educational administrative details modified as easily as income tax rates.

There would seem to be within every educational system at any one time a certain amount of 'capital' which can be expended on curriculum development. This capital will be made up of many different components: an obvious one is money, for there will be limitations on the amount of finance available to support any reforms; others are human capital (are there sufficient leaders and ancillary workers with the requisite expertise available?) and goodwill (will teachers, parents, professional mathematicians and other interested parties support the intended reforms?). If insufficient capital is available then the reforms will fail. Unfortunately, ensuring that the capital exists within the system to effect the required reforms is not easy.

It must be acknowledged also that educators are faced with a most daunting challenge for their goals can <u>never</u> be attained. Just when 'success' is in sight, new targets will be set. Criticisms of educational systems continue, often with increasing severity, but little credit is given to educators for the enormous advances made in the last half-century. In England there are now more post-graduate degrees awarded every year than there were 18+ school-leaving certificates fifty years ago; more first (bachelor) degrees than 16+ certificates. The West German educational system has been remodelled since 1950 (Max Planck Institute, 1979). Significant changes then do occur, including changes for the better, yet educational progress is not achieved overnight. In fact, a major problem for the 1990s will be:

How can politicians, administrators and educators be brought better to understand the problems and limits of educational change and to appreciate the investment that will be needed if reforms are to be successfully implemented?

A most important factor is that goals must be set which challenge, but which will not inevitably lead to failure and demoralisation. 'Change' must become more clearly linked with 'progress' than with 'upheaval'.

3. Strategies for change

Any change implies an input to and a perturbation of a system.

Where and how does one most effectively intervene in a system? What are the consequences of different types of intervention?

These two last questions force us to realise that many types of change are made/required in school mathematics. In some cases the initial move must be made at a governmental level, far away from individual schools and classrooms: above we gave two examples drawn from England and France. At the other extreme, change must be to a large extent locally directed, such as the placing of a micro in a classroom.

The point at which pressure for change is applied has great
significance so far as the chances of successful implementation are
concerned. Change requires effort and it is often accompanied by early
set-backs and doubts. If the teacher has been the initiator of change,
or has played a part in the setting of the new goals or in the
definition of new aims or practices, or has merely been responsible for
making the decision that locally that change should be made, then (s)he
is effectively a 'shareholder' in the initiative: (s)he has an obvious
incentive, which will help the 'teething' problems to be surmounted.
However, if the decision has been taken centrally - as some must - then
the position is very different. Now teachers must be convinced of the
need to change and provided with suitable support and incentives.
Remember, the teacher is being asked to desert the familiar, for the
unknown: a psychological barrier has to be surmounted.

On many occasions the strategy employed for persuading teachers
to change has been to pour scorn upon current practice - existing
procedures have been derided and innovation offered as a panacea. The
results have usually been disastrous. To deride current practice is to
discount the teacher as a judge and a professional: it is to deny to
the teacher a degree of professionalism on which any successful
innovation must build. The teacher, thereafter, is at the mercy of any
external agents, be they from higher education, industry or elsewhere,
who have firm views on 'what schools should be doing'.

Of course, any change we seek to effect should be an advance on
current practice - it should offer 'more for less' - and the innovator
must be able to explain its advantages. Yet changes must foster
teacher development and not destroy teachers' self-confidence and
spirit of professionalism.

The teacher's crucial role in educational change is becoming
increasingly accepted. However, the vast majority of the teachers of
the 1990s are already in post and have firm ideas about their role in
school and clear expectations regarding both the curriculum and their
students. Significant changes in school mathematics will only be
achieved if there are marked changes in the perceptions and attitudes
of these teachers and if they are assisted to develop necessary new
skills.

How can one attempt to change attitudes, values, skills
teaching styles, etc. and develop confidence in the use of new methods
and technology?

In some countries, it has been suggested that vast sums should
be expended in a quest for 'excellence' (see, for example, National
Science Board (1983) for a proposed plan of action). Can high
standards in education be achieved through planning and expenditure in
the same way as a space programme, and, if so, how? For what kinds of

quality can one legislate? Certainly, there are historical examples to
suggest that one can buy or legislate for enhanced competence according
to a somewhat limited range of criteria. Such questions are not merely
rhetorical; they must be seriously considered by educational ministries
and educators everywhere. Better assessment of teachers could go some
way to weeding out incompetents and so to raising base standards; but
will it bring 'excellence' in its wake?

4. Agencies for change

Various agencies have been used to help bring about change and
to improve teaching. In some countries, e.g. England, the examination
system is being used as a means of effecting change. That is, changes
are being made to the examination system not in response to happenings
in the classroom, but rather in order to reform classroom practice.
This strategy has an obvious appeal, for it demands a clear response
from all. Yet it is not without its dangers. First, if the changes
are to extend beyond mere content, then one does not circumvent the
huge problems associated with curriculum development: unless one can
deal effectively with these then a modus operandi will evolve which
satisfies the 'letter of the law' but not its spirit. Secondly, using
examinations as a means of change further elevates their role within
education, a part already over-emphasised. Thirdly, historical
precedents for the use of such strategies are not encouraging so far as
their long-term effects are concerned, even though in the short-term
there can be benefits.

Yet not to use examination systems, where they exist, would be
foolish. The changes which it is wished to effect must, however, be
supported through other agencies.

Until the 1950s, change took place slowly and within relatively
small educational systems. The main agents were individuals working
through, and together with, teacher associations. Since that time the
rate at which schools have been asked to change has increased
enormously and systems have expanded many times over. New agencies for
initiating and supporting change have arisen.

The 'curriculum development project' was an invention of the
'50s and '60s. It flourished in a variety of modes reflecting the
different degrees of influence of serving teachers (see, for example,
Howson, Keitel and Kilpatrick op.cit.), and is still an essential
weapon in the curriculum developer's armoury. Now, however, the need
for 'after-care' and for the continued support of teachers embarking on
new courses is more widely realised.

In some countries, change is encouraged and supported through
the appointment of 'advisers' or 'teacher-advisers'. A brief
description of the adviser's role in curriculum development, with
particular reference to Australia, is to be found in Howson and Malone

(1986). Certainly, advisers have an important part to play. However, their most influential role is to some degree at odds with their title: it is to be hoped that, in certain circumstances, they have good 'advice' to offer, but it is as a facilitator that their greatest contribution can be made – for example, by bringing teachers together to work on a task, or by putting a teacher faced with a particular problem in touch with others who have encountered the same problem and have come up with (perhaps differing) solutions. For it must be remembered that a variety of solutions will need to be offered for most teaching problems. Teachers have their own style of teaching and a solution which works well for one (perhaps the adviser) may fail to match the teaching style of another. Moreover, it must be realised, that good teachers will not necessarily make good advisers, and that appointing advisers will reduce the number of good teachers still in the classroom. Can an outstanding teacher make a more significant contribution to curriculum development by remaining in a school and running an excellent and paradigmatic department there?

This leads to discussion of the notion of 'lighthouse' or 'magnet' schools, centres of excellence whose examples can be emulated. In an unofficial way, this was a key form of curriculum development in the past. In almost every country one can point to certain institutions whose innovatory work was greatly to influence that of other schools for decades: 'great educators' are frequently linked with the names of the institutions in which they put their theories into practice – Basedow and the Philanthropinum, Pestalozzi at Yverdun and elsewhere, Decroly and L'Ecole de l'Ermitage. More recently, the idea has arisen of the planned creation of magnet schools in order to create a pattern which will spread throughout the system. The benefits of this approach are forcibly argued in Volume 2 of National Science Board (1983). The danger that has to be averted in this approach is that the scales are not too heavily weighted in favour of the magnet schools: the conditions in them, whether they relate to staffing, equipment or general levels of support, must not differ too much from those in ordinary schools. If they do, then the latter have a ready-made excuse for inaction, and a justifiable cause for complaint. It may be better then if centres of excellence are spotted and brought to general attention, rather than created ab initio.

Other support for teachers can come from formal networks of centres, such as the IREMs in France. Here we have agencies for in-service education and for the mediation and interpretation of meaning within a highly centralised system.

All these types of agency have a potential contribution to make to curriculum development in the 1990s. Internal constraints in a particular country may rule some of them out of consideration. Yet in every country it will be necessary to weigh up the relative efficiencies of the various agencies, bearing in mind local conditions, and to determine upon a strategy most likely to ensure improvements in mathematics education. It will be necessary for administrators and curriculum developers to do this, but it will also be necessary for teachers to formulate designs and demands in order to assist them to improve their teaching. What will best assist a teacher who wishes to change? Almost certainly the answer will hinge on 'time': time to read, to think, to plan, to discuss, to visit other schools, ... It is essential that teachers should grasp the initiative and see themselves as primary agents for change, and not tools to be manipulated by others.

Yet many teachers will not grasp the initiative and almost certainly there will be centralised attempts to raise standards, possibly by legislation. These attempts will seek to restrict the freedom of the bad teacher to carry on teaching badly – the danger that one must seek to avert is that they will not unduly restrict the freedom of the good teacher, for it is on the contributions of outstanding individuals, on 'lighthouse' teachers who point the way to others, that educational progress has always depended and will continue to depend.

5. Planning for Change

The Kuwait group began the task of identifying the elements that both demand and/or influence the need for change in mathematics education – how these are taken on board by any one nation must be essentially determined by its specific conditions. As a group we can only communicate those elements that we identified as being common to all as issues to be addressed, make informed guesses about future needs and directions for change, and alert others to the critical dimensions for change and the relationship of current information and research to identified problems.

Yet in every country there will be a need for planned change. As we have seen the things that need to be changed are essentially attitudes, practices and content. To summarise the arguments to be found in the preceding sections:

(a)   Each country needs to establish a group or groups to develop a
        plan for change that recognises:

        the need for the acquisition of new skills by individuals,
        the need to develop short, medium and long range goals,
        that planning is a continuous and iterative process,
        the need to provide motivation for people to change.

(b)   There is also a need to measure the "capital" that is available
        to support change where "capital" includes finance, talent and
        goodwill.

(c)   There is a need to develop awareness of the problems of
        educational change in politicians, administrators, teachers,
        parents, employers and the community at large.

(d)   Each country will need to develop a number of intervention
        strategies and plan an evaluation of each.

(e)   The overriding task then is to identify the elements involved in
        bringing about change in mathematics education and determining
        the inter-relationships of these elements.

Chapter 9

The Way Ahead

In this book we have set out to ask fundamental questions about school mathematics in the next decade, and about what mathematics educators in every country should be doing now to prepare for it. These questions are not asked as part of an academic game. Given careful discussion, backed by appropriate research and experimentation, answers relevant to particular countries and to particular educational systems may be found to many of them.

Just as important, many of the questions can be posed, and answered, within the context of an individual school or classroom. As we stressed in the last chapter, it is the 'lighthouse' teacher, the one who in his or her own classroom demonstrates what is possible, who can indicate the way for successful large-scale developments. Changes based on experience have a far greater chance of success than do those based on hypotheses and pipe dreams!

Certainly the answers to the questions we pose cannot be found in the literature. We have drawn attention to many references, for it is important that we are aware of what has been done in the past, and of the experience that has been accumulated. It is not the mathematics educator's job continually to reinvent the wheel. Yet, as we have stressed so often, the challenges facing us today are of a novel type: society has changed and with it both the demands made upon us as mathematics teachers and also the technology we can utilise in meeting those demands. New problems will demand new responses.

However, although the responses to the problems will be based on the work of individuals, it is only through working together, by exchanging views, criticisms, experiences and understanding, that the way ahead for educational systems will be illuminated. ICMI itself will help promote such interchanges through its four-yearly International Congresses, the meetings of such regional bodies as the Inter-American Committee for Mathematics Education and the South-East Asia Mathematical Society, and those of ICMI's affiliated international study groups and its national sub-commissions. Future ICMI studies will also look, in greater depth, at some of the issues which we have only been able to touch upon briefly in this book.

Other responses than those of ICMI will be required, however, if change is successfully to be effected in the coming decade. We hope, then, that this booklet will help in focussing the attention of individuals, institutions, associations and ministries on key issues and will prove a useful starting-point for detailed debate, experiments and critical evaluation. If so, then the main purpose of the study will have been achieved.

Bibliography

Assessment of Performance Unit (APU), 1984, A Review of Monitoring in Mathematics, D.E.S., London (2 vols).

Becker, J.R. and Barnes, M., 1986, 'Women and Mathematics' in Carss (1986), 306-313.

Belgian National Sub-Commission of ICMI, 1982, International Colloquium on Geometry Teaching, University of Mons.

Burkhardt, H., 1983, Applications in School Mathematics - the elusive El Dorado, ERIC, Ohio State University.

Buxton, L., 1981, Do you panic about mathematics?, Heinemann.

Carson, G. St. L., 1913, Essays on Mathematics Education, Ginn.

Carss, M. (ed.), 1986, Proc. ICME 5 (Adelaide, 1984), Birkhäuser.

Christiansen, B., Howson, A.G. and Otte, M. (eds.), 1986, Perspectives on Mathematics Education, Reidel.

Christiansen, B and Walther, G., 1986, 'Task and Activity' in Christiansen et al (1986), 243-308.

College Board, 1985, Academic Preparation in Mathematics, CEEB, New York.

Connell, W.F., 1980, A History of Education in the Twentieth Century, Curriculum Development Centre, Canberra.

Crosswhite, F.J. et al, 1985, SIMS: Summary Report, USA, Stipes, Ilinois.

D'Ambrosio, U., 1985, Socio-cultural bases for Mathematics Education, UNICAMP, Brazil.

   1986, 'Socio-cultural bases for Mathematics Education' in Carss (1986), pp 1-6.

Damerow, P. et al, 1986, Mathematics for all, UNESCO, Paris.

Dalin, P., 1978, Limits to Educational Change, Macmillan.

Dawe, L.C.S., 1983, 'Bilingualism and mathematical reasoning in English as a second language, Ed. Stud. Math., 14, 325-353.

Engel, A., 1977, <u>Elementarmathematik vom algorithmischen Standpunkt</u>, Klett (trs. 1984, <u>Elementary Mathematics from an Algorithmic Standpoint</u>, Keele Mathematical Education Publications, England).

Erlwanger, S.H., 1974, 'Case Studies of Children's Conceptions of Mathematics', Ph.D. thesis, University of Illinois.

Fafunwa, B., 1975, 'Education in the Mother tongue – a Nigerian experiment', <u>W. Afr. J. of Ed.</u>, 19(2), 213-227.

Fitzgerald, A., 1981, <u>Mathematics in Employment 16-18</u>, University of Bath, England.

Freudenthal, H., 1977, 'Teacher-training – an experimental philosophy', <u>Ed. Stud. Math.</u>, 8, 369-376.

            1983, <u>Didactical Phenomenology of Mathematical Structures</u>, Reidel.

Fujita, H., 1985, 'The Present State and a Proposed Reform of Mathematical Education at Senior Secondary Level in Japan', <u>J. Sci. Educ. Japan,</u> 9(2), 39-52.

Garden, R.A. and Robitaille, D.E.(eds.) in preparation, <u>SIMS Vol II: Student Achievement in Twenty Countries</u>, Pergamon Press.

Grugnetti, L., 1979, 'School mathematics makes Sardinians healthier', <u>Maths in School</u>, 8(5), 22-23.

Hart, K.M. (ed.), 1981, <u>Children's Understanding of Mathematics: 11-16</u>, John Murray.

Hembree, R. and Dessart, D.J., 1986, 'Effects of Hand-held Calculators in Pre-college Mathematics Education: a meta-analysis', <u>J. Res. Math. Ed.</u>, 17, 83-99.

Howson, A.G., Keitel, C. and Kilpatrick, J., 1981, <u>Curriculum Development in Mathematics</u>, Cambridge University Press.

Howson, A.G. and Malone, J., 1986, 'Curriculum Development' in Carss, (1986), 187-196.

Howson, A.G. and Mellin-Olsen, S., 1986, 'Social Norms and External Evaluation' in Christiansen <u>et al</u>, (1986), 1-48.

ICMI, 1978, <u>Change in Mathematics Education since the late 1950's: Ideas and Realisation</u>, Reidel.

ICMI, 1986a, The Impact of Computers and Informatics on Mathematics
        and its Teaching (Ed. R.F. Churchhouse et al), Cambridge
        University Press.

        1986b, 'Mathematics as a Service Subject' (A.G. Howson, et al),
        L'Enseignement Mathématique.

Jacobsen, E. and Dawe, L. 1986, 'Language and Mathematics' in Carss
        (1986), 261-272.

Keitel, C., 1985, 'Mathematik für alle: Pädagogisches Postulat oder
        gesellschaftliche Anforderung', Neuwe Wiskrant, 5(1), 48-59.

Kilpatrick, J., 1981, 'Research on Mathematical Learning and Thinking
        in the US', Res. en Did. des Math., 2(3) 363-379.

Mason, J., 1985, 'What happens when the machine is switched off', in
        Document de Travail: ICMI Conference, Strasbourg IREM, France.

Max Planck Institute for Human Development and Education, 1979, Das
        Bildungswesen in der Bundesrepublik Deutschland, Rowohlt (trs.,
        1983, Between Elite and Mass Education, SUNY Press, USA).

Mellin-Olsen, S. 1981, 'Instrumentalism as an educational concept',
        Ed. Stud. Maths., 12, 351-367.

National Research Council (NRC), 1985, Mathematics, Science and
        Technology Education: Research Agenda, National Academy Press,
        Washington, D.C.

National Science Board, 1983, Educating Americans for the 21st Century
        (2 vols), National Science Foundation, Washington, D.C.

Papert, S., 1980, Mindstorms, Harvester Press.

Quadling, D.A. et al, 1986, New Perspectives on the Mathematics
        Curriculum, DES, London.

Romberg, T.A. (ed.), 1984, School Mathematics: Options for the 1990s,
        ASERI, Washington, D.C.

Romberg, T.A. and Carpenter, T., 1985, 'Research on Teaching and
        Learning Mathematics: Two Disciplines of Scientific Inquiry' in
        Wittrock, M. (ed.) Handbook of Research on Teaching (3rd Edn.),
        New York.

Royal Society, 1986, Girls and Mathematics, Royal Society, London.

Steiner, H.-G. (ed.), 1980, Comparative Studies of Mathematics
        Curricula - Change and Stability, 1960-1980, Institut für
        Didaktik der Mathematik, Bielefeld.

Thom, R., 1973, 'Modern mathematics: does it exist?' in Howson, A.G.
    (ed.), Developments in Mathematics Education, Cambridge
    University Press.

Travers, K.J. et al, in press, SIMS Vol. I: Analysis of the
    International Mathematics Curriculum, Pergamon Press.

Wheeler, D.H., 1984, 'Research Problems in Mathematics Education',
    For the Learning of Mathematics, (1) 40-47, (2) 39-43, (3)
    22-29.

Wilson, B.J., 1981, Cultural Contexts of Science and Mathematics
    Education, University of Leeds, England.

Wirszup, I., 1981, 'The Soviet Challenge', Educ. Leadership, 38 (5)
    358-360.